WHAT MAKES BIOLOGY UNIQUE?

This new book, a collection of revised, collected, and some new essays written in time for his 100th birthday by the most eminent evolutionary biologist of the past century, explores biology as an autonomous science, offers insights on the history of evolutionary thought, critiques the contributions of philosophy to the science of biology, and comments on several of the major ongoing issues in evolutionary theory. Notably, Ernst Mayr explains that Darwin's theory of evolution is actually five separate theories, each with its own history, trajectory, and impact. Natural selection is a separate idea from common descent, and from geographic speciation, and so on. A number of the perennial Darwinian controversies may well have been caused by the confounding of the five separate theories into a single composite. Those interested in evolutionary theory or the philosophy and history of science will find useful ideas in this book, which should appeal to virtually anyone with a broad curiosity about biology.

Ernst Mayr is Professor Emeritus at Harvard University and former Director of the Museum of Comparative Zoology. For his contributions as an evolutionary biologist, taxonomist, and ornithologist, as well as historian and philosopher of biology, Mayr has been called "the Darwin of the 20th century." This is his twenty-fifth book.

What Makes Biology Unique?

Considerations on the autonomy of a
scientific discipline

Ernst Mayr

Harvard University

PUBLISHED BY THE PRESS SYNDICATE OF THE UNIVERSITY OF CAMBRIDGE
The Pitt Building, Trumpington Street, Cambridge, United Kingdom

CAMBRIDGE UNIVERSITY PRESS
The Edinburgh Building, Cambridge CB2 2RU, UK
40 West 20th Street, New York, NY 10011-4211, USA
477 Williamstown Road, Port Melbourne, VIC 3207, Australia
Ruiz de Alarcón 13, 28014 Madrid, Spain
Dock House, The Waterfront, Cape Town 8001, South Africa

http://www.cambridge.org

First published 2004

Printed in the United States of America

Typeface Garamond 3 11/15 pt.   *System* LATEX 2ε   [TB]

A catalog record for this book is available from the British Library.

Library of Congress Cataloging in Publication Data
Mayr, Ernst, 1904–
What makes biology unique?: considerations on the autonomy of a scientific
discipline / Ernst Mayr.
p.   cm.
Includes bibliographical references and index.
ISBN 0-521-84114-3
1. Biology – Philosophy.   2. Evolution (Biology) – Philosophy.   I. Title.
QH331.M375   2004
570'.1 – dc22   2004045888

ISBN 0 521 84114 3 hardback

To my daughters

Christa Eisabeth Menzel

and

Susanne Mayr Harrison

in love and gratitude for

all they have given me

Contents

CONTENTS

Preface

THIS WILL BE MY LAST SURVEY of controversial concepts in biology. I have previously published papers on nearly all these subjects, in some cases more than one. Indeed, an analysis of my bibliography reveals that I have discussed the species problem in no fewer than sixty-four of my publications, and have been involved in numerous controversies. What I now offer is a revised, more mature, version of my thoughts. I am not so optimistic to believe that I have settled all (or even most) of these controversies, but I do hope to have brought clarity into some rather confused issues.

What I do not understand is why most philosophers of science believe the problems of the philosophy of science can be solved by logic. Their interminable arguments, documented by whole issues of the journal *Philosophy of Science*, show that this is not the best way to reach a solution. An empirical approach (see, for example, chapter 3 for teleology and chapter 4 for reduction) seems to be a better way.

Indeed, this conclusion raises a legitimate question – whether the traditional approach of the philosophy of science is really the best possible one. This possibility must be faced if one plans to develop a philosophy of biology. The traditional approach is based on the assumption that biology is a science exactly like any of the physical sciences, but there is much evidence to question this assumption. This raises the troubling question of whether one should not choose a different approach for the construction of a philosophy of biology from the one hitherto traditional in the philosophy of science. An answer to this question requires a deep analysis of the conceptual framework of biology and its comparison with the conceptual framework of physics. Such an analysis and comparison apparently have never been made. To do that is the major objective of this work.

During this task I discovered that throughout biology there are numerous unresolved controversies dealing with problems such as the species problem, the nature of selection, the use of reduction, and several others. It is necessary to obtain clarity on these problems before one can deal with the problem of the status of biology compared with various physical sciences. Any uncertainty about some minor problem may be used by some opponents of certain major theories of biology to reject that basic theory. This has happened particularly often with Darwinism as a whole. There are still some uncertainties about some evolutionary phenomena like the conflict between the explosive speciation of cichlid fishes in the lakes of eastern Africa and the stasis of the phenotype in living fossils, but the validity of the basic Darwinian paradigm is now so firmly established that it simply cannot be questioned any longer.

However, the critical analysis of the controversial problems discussed in chapters 5–11 will help to clarify some obscure points. At first sight, bringing the topics of these chapters together would seem to produce disturbing heterogeneity. More detailed study shows, however, that the conclusions reached in each of these chapters make an important contribution to our understanding of evolution as a whole. Those who are teaching a course on the history and philosophy of biology will find the chapters on the maturation of Darwinism, on selection, and on the evolution of the human particularly helpful. These chapters also supplement treatments of these subjects in *What Evolution Is* (Mayr 2001).

LITERATURE CITED

Mayr, E. 2001. *What Evolution Is.* New York: Basic Books.

Acknowledgments

THIS WORK, the product of nearly eighty years of study, owes a great debt of gratitude to scores of friends and mentors. Most of you are no longer with us, such as Erwin Stresemann, Bernard Rensch, Theodosius Dobzhansky, Michael Lerner, James P. Chapin, J. B. S. Heldane, E. B. Ford, David Lack, Konrad Lorenz, Niko Tinbergen, and so many others.

Fortunately there are some to whom I can say my thanks in person. Walter Bock is the one to whom I owe the most. He has read my draft manuscripts critically and I have greatly benefited from his constructive suggestions. I frequently consulted Francisco Ayala, Jared Diamon, Doug Futuyma, Michael Ghiselen, Verne Grant, Axel Meyer, David Pilbeam, Frank Sulloway, and Bruce Wallace and always received useful information and constructive advice. I also consulted Fred Burkhardt, J. Cain, J. Coyne, James Crow, Frances DeWaal, J. Haffer, François Jacob, Lynn Margulis, Robert May, Eviatar Nevo, J. W. Schoff, Steve Stanley, Robert Triver, James D. Watson, E. O. Wilson, and R. W. Wrangham. They contributed valuable information and wise counsel. All this improved the reliability and competence of this volume.

I had no permanent secretary in the period when this volume was in production, but I was provided with part-time student help during the last seven years, who made invaluable contributions to the quality of this work. But it was Alison Pine who very efficiently saw this volume through the press in the last two years. To her, I owe a special debt of gratitude.

Introduction

MY FATHER HAD A LARGE LIBRARY. Even though he was a jurist by profession, his major interests were history and philosophy, particularly the German philosophers Kant, Schopenhauer, and Nietzsche. I never read any of these philosophy books, unless one classifies Haeckel (Welträtsel) as a philosopher. However, in my parents' home philosophy was always referred to with great respect. Philosophy was the favorite reading of my father's maiden sister whom the family considered brilliant.

My real contact with philosophy, however, did not come until I prepared myself for the philosophy portion of my PhD examination. At the University of Berlin, one had to pass an examination in philosophy to complete a PhD. I took courses in the history of philosophy and a seminar in Kant's *Critique of Pure Reason*. Frankly, I really did not understand what it all was about. I was permitted to specify in what branch of philosophy

I wanted to be examined and I was duly examined in positivism as I had specified. I passed with an A because I had been well prepared.

As a result of my studies, I concluded that the traditional philosophy of science had little if anything to do with biology. When I inquired (ca. 1926) which philosophers would be most helpful to a biologist, I was told Driesch and Bergson. When I left for New Guinea one and a half years later, the major books of these two authors were the only books I dragged around with me in the tropics for two and a half years. In the evenings, when I was not busy with bird skinning, I would read in these two volumes. As a result, by the time I returned to Germany, I had concluded that neither Driesch nor Bergson was the answer to my search. Both authors were vitalists and I had no use for a philosophy based on such an occult force as the *vis vitalis*.

But I was equally disappointed by the traditional philosophy of science, which was all based on logic, mathematics, and the physical sciences, and had adopted Descartes' conclusion that an organism was nothing but a machine. This Cartesianism left me completely dissatisfied and so did saltationism. Where else could I turn?

For the next twenty years or so, I more or less ignored philosophy, but then, in due time, my activities in theoretical systematics and even more so in evolutionary biology brought me back to philosophy. I developed a vague feeling that the new concepts and principles encountered in the more theoretical branches of biology might be a good starting point for a genuine philosophy of biology. But here I had to be very careful. I did not want to fall into a trap like vitalism or become a teleologist, like Kant in his *Critique of Judgment*. I was determined not to accept any principles or causes that were in conflict with the Newtonian natural laws. The biology for which I wanted to find a philosophy had to qualify as a genuine, bona fide science.

Even though quite a few books were published in the twentieth century entitled *The Philosophy of Biology*, they live up to this title only in part. Works such as those of Ruse (1973), Kitcher (1984), Rosenberg (1985), and Sober (1993) deal with biological issues and theories but use the same epistemological framework as books on the philosophy of physics. One looks in vain for an adequate treatment of the autonomous aspects of biology, such as biopopulations and dual causation (explanation). Even though much of the methodologies of the philosophy of physical sciences can also be used in a philosophy of biology, the neglect of the specifically biological subjects leaves a painful gap. Owing to their basic philosophy, these volumes have been referred to as Cartesian. Those who were looking for a philosophy of biology had the choice between a volume that was either vitalistic in its basic spirit or Cartesian.

I had a half-hearted ambition to write a book that would fill the gap, but I realized that I was deficient in my background in philosophy. Also I was still preoccupied with unfinished researches in systematics, evolution, biogeography, and the history of biology. I simply was not in a position to try to compose such a philosophy of biology as I had in mind.

What I could do instead was to write a series of essays that might serve as a basis for such a book by a properly qualified philosopher. I have written such essays for the last twenty years; sometimes an earlier version was replaced in due time with a more mature one. Indeed, of the twelve chapters in this volume, all but four (chapters 1, 4, 6, and 10) are considerably revised versions of earlier publications. A reader casting a quick glance at the list of the chapter headings might come to the wrong conclusion that this book is a hodgepodge of unrelated themes. But this is not the case as I will now describe in a short characterization of each chapter.

The historian of biology is in a peculiar predicament. There were a number of fields dealing with the living world – physiology, taxonomy, and medicine-related embryology – in which studies were done that later became respectable components of the biological sciences. But in the eighteenth and early nineteenth centuries, they were not treated as part of the cohesive science eventually recognized as biology.

Even though Linnaeus led to a great flourishing of systematics, it was really Buffon (Roger 1997) who directed attention to the living organism. The word biology was introduced around 1800 independently by three authors, but it described something that was going to come and not a field that already existed. It finally came in the nineteenth century when in a period of about forty years all the major subdivisions of biology were established. These developments are indicated by the following names and dates: K. E. von Baer (1828), embryology; Schwann and Schleiden (1838–1839), cytology; J. Müller and Bernard (1840s–1850s), physiology; Darwin and Wallace (1858–1859), evolution; and Mendel (1866, 1900), genetics. Biology developed into a separate branch of science during this forty-year period. But it was not until the second half of the twentieth century that biology acquired dominance among the sciences.

The object of each chapter

Chapter 1 – Science and sciences

In Chapter 1 I show that biology is a bona fide science, even though it has some properties that are not found in the physical sciences. What is important, however, is that biology has the indispensable characteristics of true sciences such as chemistry and physics. It is justified to try to develop a branch of the philosophy of science devoted to biology.

Chapter 2 – The autonomy of biology

However, I also found that biology, even though it is a genuine science, has certain characteristics not found in other sciences; in other words, I show in this chapter that biology is an autonomous science.

The remaining ten chapters discuss various aspects of biology that must be fully understood by anyone wanting to study the philosophy of biology. The conclusions reached in these chapters will strengthen the foundations of a genuine philosophy of biology.

Chapter 3 – Teleology

Biology could not be accepted as a bona fide science until it eliminated cosmic teleology from its framework of theories. Therefore, it is essential to show that the word teleological has been used for five different kinds of phenomena or processes in nature, which must be carefully distinguished from each other. Satisfactory empirical explanations are available for four kinds of phenomena or processes that traditionally are referred to as teleological; these can be explained exhaustively by natural laws. Yet no evidence has ever been found for the existence of the fifth one, cosmic teleology.

Chapter 4 – Analysis or reductionism?

Until the middle of the twentieth century, an important philosophical belief of the physicalists was that a phenomenon had to be reduced to its smallest components to achieve a complete explanation. This was generally interpreted as meaning that explanation could be achieved only at the lowest level of organization. This conclusion was particularly disturbing for biologists, because at the lowest levels of organization such a reduction abandoned biology and dealt exclusively with physical phenomena. However, I will show in this chapter that such reduction is not only not necessary but indeed quite impossible. The support for reduction was in part the result of a confusion with the process of analysis. Analysis is and always will be an important methodology in the study of complex systems. Reduction, on the other hand, is based on invalid assumptions and should be removed from the vocabulary of science.

Chapter 5 – Darwin's influence on modern thought

Charles Darwin contributed many of the concepts on which the paradigm of modern biology rests. Some of them were controversial for a long time and are still opposed by certain evolutionists. A full understanding of the autonomy of biology therefore is not possible without an analysis of Darwinism. Indeed, modern biology is conceptually Darwinian to a large extent. Although I attempted in previous publications to characterize this Darwinian contribution to our modern biological thinking, its importance for the philosophy of biology is so great that this renewed analysis should be welcome.

Chapter 6 – Darwin's five theories of evolution

Darwin, throughout his life, referred to his theorizing on evolution as "my theory," in the singular. However, it is now quite clear that Darwin's evolutionary paradigm consists of five theories, which are independent of each other. Failing to appreciate this independence unfortunately led Darwin, and others who followed him, to several misinterpretations. One will never fully understand the autonomy of biology if one does not understand the nature of Darwin's five theories.

Chapter 7 – Maturation of Darwinism

The set of ideas and theories that leading evolutionists now consider to be the basic components of Darwinism are still remarkably similar to Darwin's original proposals in 1859 – largely but not entirely. In particular, Darwin had not realized that "his theory" [in the singular] is actually a compound of five different theories. These were accepted by other evolutionists at different times, with natural selection, after nearly eighty years of argument, accepted as the last.

The acceptance of evolution is of course a prerequisite for acceptance of the other four theories. But the validity of each of these four theories is independent of the validity of the other three. One can have a theory of speciation even if one rejects natural selection or gradualism. Many of the Darwinian controversies were due to the neglect of the finding that the validity of each of the four Darwinian theories is largely independent of the validity of the others.

Chapter 8 – Selection

This theory (or bundle of theories) was, for several reasons, longest resisted. Indeed, our modern concept of this theory differs in a number of ways from the original Darwinian version. For instance, we now consider selection more as a process of nonrandom elimination than of positive selection, and this may facilitate the survival of more and more deviant variants. Also, we no longer consider variation and elimination simply as each other's opposites but are beginning to consider the production of variety and the succeeding step of elimination as two steps in a single process. There remains considerable uncertainty about the role of variation during the evolutionary process, but there is no argument that selection is involved in nearly every instance of evolutionary change. A knowledge of all aspects of selection therefore is basic for a full understanding of evolution.

Chapter 9 – Do Thomas Kuhn's scientific revolutions take place?

It is extraordinary how biology has changed in the last two hundred years: first the establishment of biology as a valid science in the years from 1828 to 1866, then the Darwinian revolution, then genetics and the new systematics, and finally the revolution of molecular biology. The philosopher is deeply interested in the nature of these changes. Were they gradual or did they occur in a number of scientific revolutions, and if so what was the nature of these revolutions? One cannot understand the nature of the currently accepted science of biology unless one understands the nature of the conceptual changes of the last two hundred years. In particular, I attempt to answer in this chapter the question of whether Kuhn's concept of scientific revolutions is or is not supported by biology.

Chapter 10 – Another look at the species problem

No matter in what branch of biology one is interested, it is necessary to work with species. This is the major unit in biogeography, in taxonomy, and in all comparative branches of biology. Evolution is characterized by irreversible changes at the species level. Considering this outstanding importance of the species in biology, it strikes me as almost scandalous that there is still so much disagreement and uncertainty about almost every aspect of species. There is no other problem in biology on which more has been written in recent years and less unity has been achieved than the species problem. Any discussion of the autonomy of biology that did not attempt to shed light on the origin and the nature of species would be incomplete. My own account focuses on the reasons for this long-standing and seemingly insoluble problem and makes suggestions for a solution.

Chapter 11 – The origin of humans

It was one of Darwin's shocking findings that the human species is not something altogether different from the rest of the living world, as nearly everybody believed, but instead is part of it – indeed that apes are the ancestors of humans. Even though this conclusion had already been made inevitable, on the basis of both comparative biology and the fossil record, it has now been a thousandfold confirmed by molecular biology. What is particularly interesting is that by proposing historical narratives including the life history of our ancestors, it is possible to reconstruct a rather convincing hominid history. The scenario suggested in this chapter is based largely on inferences, but they can be tested against a great deal of evidence from fossils and from molecular biology. The novel historical narrative suggested by me will have to be tested again and again. However, it has the great advantage that it provides a cohesive and quite plausible account of the various stages by which a chimpanzee-like ancestor in the rainforest evolved into *Homo sapiens*. It is precisely the autonomous features of biology that make a plausible reconstruction possible. It produces a solid foundation for the reconstruction of human history, which a purely physics-based explanation would never be able to provide.

Chapter 12 — Are we alone in this vast universe?

This question has been asked for more than two thousand years. As an outgrowth of space research in recent years, a definite research program has been developed, trying to establish contact with any possible civilizations elsewhere in the universe. Those who have given thought to this project can readily be assigned to two groups: an optimistic one consisting almost entirely of physical scientists, particularly astronomers. They are convinced that a search for extraterrestrial intelligence is promising. By contrast a pessimistic group, consisting mostly of biologists, has developed a list of reasons why such a search is totally hopeless. In this chapter, I present the biological reasons, usually neglected by astronomers, why there is such a low probability of success.

LITERATURE CITED

Kant, I. 1781. *Critique of Pure Reason.*

Kitcher, P. 1984. 1953 and all that. *Philosophical Reviews*, 93:335–373.

Rosenberg, A. 1985. *The Structure of Biological Science.* Cambridge: Cambridge University Press.

Ruse, M. 1973. *The Philosophy of Biology.* London: Hutchinson.

Sober, E. 1993. *Philosophy of Biology.* Boulder: West View Press.

I

Science and Sciences

BIOLOGY IS A SCIENCE; there is no argument about this statement, or is there? Doubts about this claim have been suggested by important differences among various widely accepted definitions of science. A comprehensive, pragmatic definition of science might be "Science is the human endeavor to achieve a better understanding of the world by observation, comparison, experiment, analysis, synthesis, and conceptualization." Another definition might be "Science is a body of facts ('knowledge') and the concepts that permit explaining these facts," and there are numerous others. In a recent book (Mayr 1997:24–44) I have devoted a twenty-page chapter to a discussion of the question, "What is science?"

Difficulties arise because the term science also has been used for so many human activities beyond the natural sciences, such as the social sciences, political science, military science, and more distant areas such as Marxist science, Western science, feminist science, and even Christian Science and Creationist Science. In all these combinations, the word science is used in a misleadingly inclusive sense. Equally misleading, however, is the opposite extreme, the decision of some physicists and physicalist philosophers to restrict the word science to mathematically based physics. A vast literature shows how difficult, indeed impossible, it seems to be to draw a line between incontrovertible science and adjacent fields. This diversity is a heritage of history.

One can claim that science originated in preliterary times when people began to ask "how?" and "why?" questions about the world. Much of what philosophers were doing in Greece and the Ionian colonies in Asia Minor and southern Italy was rudimentary science. Aristotle's work was a very respectable beginning of the science of biology. However, it is rather generally accepted that the so-called scientific revolution of the sixteenth and seventeenth centuries, characterized by Galileo, Descartes, and Newton, was the real beginning of what is now called science. At that time most phenomena in the inanimate and the living world were not yet explained in terms of natural causes, and God was still considered the ultimate cause of everything. However, in due time secular explanations were ever more widely adopted and considered legitimate science. It dealt primarily with two branches of science, mechanics and astronomy. Not surprisingly, at that time the concept of science was the concept of these two physical sciences. For Galileo, mechanics was the dominant science and it remained so for hundreds of years.

When intellectual life revived after the Middle Ages, there was no word for what we now call science. Indeed, the English word science for what modern people call science was introduced by Whewell as late as 1840. However, at the time of the scientific revolution in the sixteenth, seventeenth, and eighteenth centuries, science was conceived very broadly by some authors and very narrowly by others.

The philosopher Leibniz was exemplary for the broad conception. For him and his followers, a "science was a body of doctrine that could be known systematically and with a high degree of certainty; it was contrasted with 'opinion,' that which could only be understood with a lesser degree of certainty, or with 'art,' that which involved a practice rather than doctrine" (Garber and Ariew 1998). Science, thus conceived, included natural sciences, natural history (including medicine, geology, and chemistry), mathematics, metaphysics and even theological writings, European history, and linguistics. It is this exceedingly broad conceptualization of science that is still alive in the German concept of the *Geisteswissenschaften*. Everything that in Anglophone countries is included in the humanities is referred to in the German literature as *Geisteswissenschaften*.

This includes the study of classics, philosophy, linguistics, and history. As a result one recognized in Germany two kinds of *Wissenshaften*, the natural sciences and the *Geisteswissenschaften*. There is indeed some justification to include some of the cited disciplines of the humanities among the real sciences. They employ methods and adopt principles that are analogous to those of the natural sciences. This led to an argument about where to draw the line between the two kinds of sciences. Considering how similar evolutionary biology is to historical science and how different it is from physics in conceptualization and methodology, it is not surprising that drawing a definite line between the natural sciences and the humanities is so difficult, indeed nearly impossible. For example, someone might place this line between functional and evolutionary biology, attaching functional biology to the natural sciences and evolutionary biology to the science of history.

Physicalism

One extreme is Galileo's (1564–1642) science. At his time only one science existed, that of mechanics (including astronomy). Hence, when Galileo characterized science, he based it on his knowledge of mechanics.

Having no other sciences to compare mechanics with, he did not realize that his characterization of "science" (= mechanics) included two quite different sets of characteristics – those true for any genuine science and those true only for mechanics. For instance, he did not realize that mathematics plays a far greater role in mechanics than in most other sciences. Hence, mathematics played a dominant role in Galileo's image of science. He insisted that the book of nature "cannot be understood unless one first learns to comprehend the language and read the letters in which it is composed. It is written in the language of mathematics and its characters are triangles, circles, and other geometric figures without which it is humanly impossible to understand a single word of it; without these one wanders about in a dark labyrinth" (Galileo 1632). Nor, quite naturally, was any discrimination made by anyone else, because at first there were no other sciences with which to compare mechanics. Physics with a mathematical foundation became the exemplar of science for Galileo, Newton, and all the other greats of the Scientific Revolution. This physicalist interpretation dominated the thinking of the philosophers of science. And this remained so for the next three hundred fifty years. Curiously, it was quite generally ignored in discussions of science in those centuries that there were now also other sciences. Instead, these other sciences were squeezed into the conceptual framework of physics. Mathematics remained the earmark of true science. Kant certified this opinion by saying "there is only that much genuine [*richtig*] science in any science, as it contains mathematics." And this greatly exaggerated evaluation of physics and mathematics has dominated science until the present day. What would be the scientific status of Darwin's *Origin of Species* (1859), which contains not a single mathematical formula and only a single phylogenetic diagram (not a geometric figure) if Kant had been right? And this was also the philosophy of science of the leading philosophers (e.g., Whewell, Herschel) that affected Darwin's thought (Ruse 1979). Yet several recent philosophers of science have published a *Philosophy of Biology* strictly based on the conceptual framework of the classical physical sciences (e.g., Kitcher 1984, Ruse 1973, Rosenberg 1985) while ignoring the autonomous aspects of biology (chapter 2).

Yes, God was the creator of this world and either directly or through his laws he was responsible for everything that existed and occurred. Science for Galileo and his followers was not an alternative to religion but an inseparable part of it, and this remained true from the sixteenth century to the first half of the nineteenth century and was accepted by the great philosophers of that period including Kant. Yet the vigorously expanding science of the eighteenth and early nineteenth centuries was able to find a natural explanation for one phenomenon after the other that previously had required invoking God's presence. Eventually, only lip service was paid to Galileo's claim about the dominant role of mathematics in science.

Even after physicalism was considerably liberalized in the last one hundred years, it remains questionable how sound a basis for a philosophy of biology it can provide. Historians of physics traditionally have exaggerated the importance of the great discoveries in physics in the 1920s (quantum mechanics, relativity, elementary particle physics, etc.). The historian Pais said, for instance, that Einstein's theories "have profoundly changed the way modern men and woman think about the phenomena of inanimate nature." But on second thought he realized his exaggeration and corrected this claim to "It would actually be better to say 'modern scientists' than 'modern men and women.'" Actually, it would be even better to say "physical scientists," because Einstein's theories did not affect other scientists at all. Indeed, to appreciate Einstein's contributions in their fullness, one needs to be schooled in the physicist's style of thinking and in special branches of mathematics. It requires much optimism to guess that even one in every 100,000 humans alive today has any insight into what Einstein's relativity is all about. Indeed, hardly any of the great discoveries in physics in the 1920s had any apparent effect on biology at all.

A proliferation of sciences

Beginning with the sixteenth century, the scientific revolution was accompanied by the origin of several other sciences, which included historical sciences such as cosmology and geology and various fields traditionally considered parts of the humanities, such as psychology, anthropology, linguistics, philology, and history. They all became increasingly scientific in the ensuing centuries. This was particularly true for research eventually combined under the name biology.

Aristotle in the fourth century B.C. had produced a remarkable contribution to biology, particularly to its methodology and principles. Even though a few additional interesting discoveries were later made in the Hellenic period and by Galen and his school, biology remained more or less dormant until the sixteenth century. Some contributions, however, were made in two widely distant areas. The medical schools from the sixteenth century on were beginning to make advances in anatomy, embryology, and physiology; at the same time, natural history, in the broadest sense of the word, was equally furthered by natural theologians like Ray, Derham, and Paley; by naturalists like Buffon and Linnaeus; and by numerous lay naturalists.

As we shall see, in the seventeenth and eighteenth centuries students of the living world, both at the medical schools and among the natural historians (natural theology), actively laid a foundation for a science of biology. Yet, that such a field as biology existed was almost universally ignored by historians and philosophers. When Kant (1790), in his *Critique of Judgment*, was quite unsuccessful in explaining the phenomena of the living world with the help of Newtonian laws and principles, he solved his dilemma by ascribing biological processes to teleology. Most other philosophers simply ignored the existence of biology. Science is physics, they said simply. More recently, the writings of philosophers of science from the Vienna School to Hempel and Nagel and to Popper and Kuhn were strictly based on and applicable to the physical sciences. When C. P. Snow decried the gap between science and the humanities, he actually described the gap between the physical sciences and the humanities. Biology was nowhere referred to in his discussions. As late as the 1970s and 1980s various philosophers (e.g., Hull 1974, Ruse 1973, Sober 1993) wrote philosophies of biology essentially based on the conceptual framework of the physical sciences. Of course, their education usually had been in logic or mathematics, rather than in biology.

Some authors broke away from this monopoly of the physical sciences (often referred to as Cartesianism) because they realized that these strictly physicalist endeavors were not an adequate foundation for a philosophy of biology. But their proposal was not the sought-for solution either, because they invoked occult forces (vitalism and teleology). The last well-known representatives of this vitalistic approach were Bergson (1911) and Driesch (1899) (see chapter 2). Even though these authors sensed that vitalism was an invalid approach, they were unable to find a better solution. For me it became clear in the 1950s that any approach to a philosophy of biology, essentially based on logic and mathematics rather than on the specifically unique concepts of biology, would be unsatisfactory. The solution had to come from biology, but what would biology have to do to find it?

Why is biology different?

In spite of spectacular developments such as genetics, Darwinism, and molecular biology, biology continued to be treated as a branch of physicalist science. Only a few philosophers realized that mechanics as well as all post-Galileian sciences, consisted of two types of attributes. These are, first, the characteristics all genuine sciences share, including the organization and classification of knowledge on the basis of explanatory principles (Mayr 1997). The other attributes consist of characteristics that are specific for a particular branch of science or group of sciences. In the case of mechanics, this would include the special role of mathematics, the foundation of its theories on natural laws, and a much greater tendency toward determinism, to typological thinking, and to reductionism than found in biology. None of these mechanics-specific characteristics plays a major role in theory formation in biology.

When the philosophy of science began to originate, the philosophers apparently took it for granted that all kinds of science were equivalent as far as philosophy is concerned. This is why Galileo, Kant, and indeed most philosophers of science applied to biology, without change, a philosophy that had been developed on the basis of mechanics. And the same guideline was used for all sciences: anthropology, psychology, sociology, and others. What is needed instead is a careful analysis of each science to determine whether its basic principles and components are adequately covered by the explanations of mechanics and more broadly by those of physics. As a first contribution to this project, I have undertaken this task for biology. My findings are presented in chapter 2, "The Autonomy of Biology."

LITERATURE CITED

Bergson, H. 1911. *L'Evolution Créatrice*. Paris: Alcan.

Darwin, C. 1859. *On the Origin of Species by Means of Natural Selection or the Preservation of Favored Races in the Struggle for Life*. London: John Murray [1964, Facsimile of the First Edition; Cambridge, MA: Harvard University Press].

Driesch, H. 1899. *Philosophie des Organischen*. Leipzig: Quelle und Meyer.

Galileo, G. 1632 (2001). *Dialogue Concerning the Two Chief World Systems. Ptolemaic and Copernican*. Translated by S. Drake. New York: Modern Library.

Garber, D., and A. Ariew. 1998. Introduction: Leibniz and the sciences. *Perspectives on Science*, 6:1–5.

Hull, D. L. 1974. *Philosophy of Biological Science*. Englewood Cliffs, NJ: Prentice-Hall.

Kant, I. 1790. *Die Kritik der Urteilskraft*. Berlin: Georg Reimer.

Kitcher, P. 1984. 1953 and all that. *Philosophical Reviews*, 93:335–373.

Mayr, E. 1997. *This Is Biology. The Science of the Living World*. Cambridge, MA: Harvard University Press, chapt. 2, pp. 24–49.

Pais, A. 1982. *Subtle is the Lord: The Science and the Life of Albert Einstein*. Oxford: Oxford University Press.

Roger, J. 1997. *Buffon: A Life in Natural History*. Translated by S. L. Bonnefoi. Ithaca: Cornell University Press.

Rosenberg, A. 1985. *The Structure of Biological Science*. Cambridge, NY: Cambridge University Press.

Ruse, M. 1973. *The Philosophy of Biology*. London: Hutchinson.

Ruse, M. 1979. *Darwinian Revolution: Science Red in Tooth and Claw*. Chicago: University of Chicago Press.

Sober, E. 1993. *Philosophy of Biology*. Boulder: West View Press.

Woese, C. R. 2002. On the evolution of cells. *Proceedings of the National Academy of Sciences*, 99:8742–8747.

2

The Autonomy of Biology

I‌T TOOK MORE THAN TWO HUNDRED YEARS and the occurrence of three sets of events before a separate science of the living world – biology – was recognized. As I will show, one can assign these events to three different sets: (A) the refutation of certain erroneous principles, (B) the demonstration that certain basic principles of physics cannot be applied to biology, and (C) the realization of the uniqueness of certain basic principles of biology that are not applicable to the inanimate world. This chapter is devoted to an analysis of these three sets of developments. This has to be done before one can accept the view of an autonomy of biology. For an earlier support of the autonomy of biology see Ayala (1968).

The refutation of certain erroneous basic assumptions

Under this heading, I deal with the support for certain basic ontological principles that later were shown to be erroneous. Biology could not be recognized as a science of the same rank as physics as long as most biologists accepted certain basic explanatory principles not supported by the laws of the physical sciences and eventually found to be invalid. The two major principles here involved are *vitalism* and a belief in cosmic *teleology*. As soon as it had been demonstrated that these two principles are invalid and, more broadly, that none of the phenomena of the living world is in conflict with the natural laws of the physicalists, there was no longer any reason for not recognizing biology as a legitimate autonomous science equivalent to physics.

Vitalism

The nature of life, the property of being living, has always been a puzzle for philosophers. Descartes tried to solve it by simply ignoring it. An organism is really nothing but a machine, he said. And other philosophers, particularly those with a background in mathematics, logic, physics, and chemistry, tended to follow him and operated as if there were no difference between living and inanimate matter. But this did not satisfy most naturalists. They were convinced that in a living organism certain forces are active that do not exist in inanimate nature. They concluded that, just as the motion of planets and stars is controlled by an occult, invisible force called gravitation by Newton, the movements and other manifestations of life in organisms are controlled by an invisible force, Lebenskraft or *vis vitalis*. Those who believed in such a force were called vitalists.

Vitalism was popular from the early seventeenth century to the early twentieth century. It was a natural reaction to the crass mechanism of Descartes. Henri Bergson (1859–1941) and Hans Driesch (1867–1941) were prominent vitalists in the early twentieth century. The end of vitalism came when it no longer could find any supporters. Two causes were largely responsible for this: first, the failure of literally thousands of unsuccessful experiments conducted to demonstrate the existence of a Lebenskraft; second, the realization that the new biology, with the methods of genetics and molecular biology, was able to solve all the problems for which scientists traditionally had invoked the Lebenskraft. In other words, the proposal of a Lebenskraft had simply become unnecessary.

It would be ahistorical to ridicule vitalism. When one reads the writings of some of the leading vitalists like Driesch, one is forced to agree with him that many of the basic problems of biology simply cannot be solved by Cartesian philosophy, in which the organism is considered nothing but a machine. The developmental biologists, in particular, asked some very challenging questions. For example, how can a machine regenerate lost parts, as many kinds of organisms are able to do? How can a machine replicate itself? How can two machines fuse into a single one like the fusion of two gametes to produce a zygote?

The critical logic of the vitalists was impeccable. But all their efforts to find a scientific answer to the so-called vitalistic phenomena were failures. Generations of vitalists labored in vain to find a scientific explanation for the Lebenskraft until it finally became quite clear that such a force simply does not exist. That was the end of vitalism.

Teleology

Teleology is the second invalid principle that had to be eliminated from biology before it qualified as a science equivalent to physics. Teleology deals with the explanation of natural processes that seem to lead automatically to a definite end or goal. To explain the development of the fertilized egg to the adult of a given species, Aristotle invoked a fourth cause, the *causa finalis*. Eventually, one invoked this cause for all phenomena in the cosmos that led to an end or goal. Kant in his *Critique of Judgment* at first tried to explain the biological world in terms of Newtonian natural laws but was completely unsuccessful in this endeavor. Frustrated, he ascribed all *Zweckmässigkeit* (adaptedness) to teleology. This was, of course, no solution. A widely supported school of evolutionists, for instance, the so-called orthogenesists, invoked teleology to explain all progressive evolutionary phenomena. They believed that in living nature there is an intrinsic striving ("orthogenesis") toward perfection. Here belongs also Lamarck's theory of evolution, and orthogenesis had many followers before the evolutionary synthesis. Alas, no evidence for the existence of such a teleological principle could ever be found, and the discoveries of genetics and paleontology eventually totally discredited cosmic teleology. For a more detailed discussion of teleology see chapter 3.

What is biology?

When we try to answer this question, we find that biology actually consists of two rather different fields, mechanistic (functional) biology and historical biology. Functional biology deals with the physiology of all activities of living organisms, particularly with all cellular processes, including those of the genome. These functional processes ultimately can be explained purely mechanistically by chemistry and physics.

The other branch of biology is *historical biology*. A knowledge of history is not needed for the explanation of a purely functional process. However, it is indispensable for the explanation of all aspects of the living world that involve the dimension of historical time – in other words, as we now know, all aspects dealing with evolution. This field is evolutionary biology.

The two fields of biology also differ in the nature of the most frequently asked questions. To be sure, in both fields one asks "what?" questions to get the facts needed for further analysis. The most frequently asked question in functional biology, however, is "how?"; in evolutionary biology "why?" is the most frequently asked question. This difference is not complete because in evolutionary biology one also occasionally asks "how" questions – for instance, how do species multiply? However, as we will see, to obtain its answers, particularly in cases in which experiments are inappropriate, evolutionary biology has developed its own methodology, that of *historical narratives* (tentative scenarios).

To truly appreciate the nature of biology one must know the remarkable difference between these two branches of biology. Indeed, some of the most decisive differences between the physical sciences and biology are true for only one of these branches, for evolutionary biology.

The emergence of modern biology

The two-hundred-year period from about 1730 to 1930 witnessed a radical change in the conceptual framework of biology. The period from 1828 to 1866 was particularly innovative. Within these thirty-eight years, both branches of modern biology – functional and evolutionary biology – were established. Yet biology was still largely ignored by the philosophers of science from Carnap, Hempel, Nagel, and Popper to Kuhn. Biologists, even though they now rejected vitalism and cosmic teleology, were unhappy with a purely mechanistic (Cartesian) philosophy of biology. But all endeavors to escape from this dilemma – such as, for example, the writings of Jonas, Portmann, von Uexküll, and several others – invariably invoked some nonmechanical forces that were not acceptable to most biologists. The solution had to satisfy two demands: it had to be completely compatible with the natural laws of the physicists, and no solution was acceptable that would invoke any occult forces. It was not until almost the middle of the twentieth century that it became evident that a solution could not be found by a philosopher who did not have a background in biology. But no such philosopher made the attempt.

It turned out that to develop an autonomous science of biology one had to do two further things. First, one had to undertake a critical analysis of the conceptual framework of the physical sciences. This revealed that some of the basic principles of the physical sciences are simply not applicable to biology. They had to be eliminated and replaced by principles pertinent to biology. Second, it was necessary to investigate whether biology is based on certain additional principles that are inapplicable to inanimate matter. This required a restructuring of the conceptual world of science that was far more fundamental than anyone had imagined at that time. It became apparent that the publication in 1859 of Darwin's *Origin of Species* was really the beginning of an intellectual revolution that ultimately resulted in the establishment of biology as an autonomous science.

Physicalist ideas not applicable to biology

Darwin's ideas were particularly important in the discovery that a number of basic concepts of the physical sciences, which up to the middle of the nineteenth century were also widely held by most biologists, are not applicable to biology. I will now discuss four of these basic physicalist concepts for which it had to be demonstrated that they are not applicable to biology before it was realized how different biology is from the physical sciences.

1. ESSENTIALISM (TYPOLOGY). From the Pythagoreans and Plato on, the traditional concept of the diversity of the world was that it consisted of a limited number of sharply delimited and unchanging *eide* or essences. This viewpoint was called typology or essentialism. The seemingly endless variety of phenomena, it was said, actually consisted of a limited number of natural kinds (essences or types), each forming a class. The members of each class were thought to be identical, constant, and sharply separated from the members of any other essence. Therefore,

variation was nonessential and accidental. The essentialists illustrated this concept by the example of the triangle. All triangles have the same fundamental characteristics and are sharply delimited against quadrangles or any other geometric figure. An intermediate between a triangle and a quadrangle is inconceivable.

Typological thinking, therefore, is unable to accommodate variation and has given rise to a misleading conception of human races. Caucasians, Africans, Asians, and Inuits are types for a typologist that differ conspicuously from other human ethnic groups and are sharply separated from them. This mode of thinking leads to racism. Darwin completely rejected typological thinking and instead used an entirely different concept, now called *population thinking* (see below).

2. DETERMINISM. One of the consequences of the acceptance of deterministic Newtonian laws was that it left no room for variation or chance events. The famous French mathematician and physicist Laplace boasted that a complete knowledge of the current world and all its processes would enable him to predict the future to infinity. Even the physicists soon discovered the occurrence of enough randomness and contingencies to refute the validity of Laplace's boast. The refutation of strict determinism and of the possibility of absolute prediction freed the way for the study of variation and of chance phenomena, so important in biology.

3. REDUCTIONISM. Most physicalists were reductionists. They claimed that the problem of the explanation of a system was resolved in principle as soon as the system had been reduced to its smallest components. As soon as one had completed the inventory of these components and had determined the function of each one of them, they claimed, it would be an easy task also to explain everything observed at the higher levels of organization. See chapter 4 for a detailed study of the validity of reductionism.

4. THE ABSENCE OF UNIVERSAL NATURAL LAWS IN BIOLOGY. The philosophers of logical positivism, and indeed all philosophers with a background in physics and mathematics, base their theories on natural laws and such theories are therefore usually strictly deterministic. In biology there are also regularities, but various authors (Smart 1963, Beatty 1995) severely question whether these are the same as the natural laws of the physical sciences. There is no consensus yet in the answer to this controversy. Laws certainly play a rather small role in theory construction in biology. The major reason for the lesser importance of laws in biological theory formation is perhaps the greater role played in biological systems by chance and randomness. Other reasons for the small role of laws are the uniqueness of a high percentage of phenomena in living systems as well as the historical nature of events.

Owing to the probabilistic nature of most generalizations in evolutionary biology, it is impossible to apply Popper's method of falsification for theory testing because a particular case of a seeming refutation of a certain law may not be anything but an exception, as are common in biology. Most theories in biology are based not on laws but on concepts. Examples of such concepts are, for instance, selection, speciation, phylogeny, competition, population, imprinting, adaptedness, biodiversity, development, ecosystem, and function.

The inapplicability to biology of these four principles that are so basic in the physical sciences has contributed a great deal to the insight that biology is not the same as physics. To get rid of these inappropriate ideas was the first, and perhaps the hardest, step in developing a sound philosophy of biology.

Autonomous characteristics of biology

The last step in the development of the autonomy of biology was the discovery of a number of biology-specific concepts or principles.

The complexity of living systems

There are no inanimate systems in the mesocosmos that are even any-where near as complex as the biological systems of the macromolecules and cells. These systems are rich in emergent properties because forever new groups of properties emerge at every level of integration. An analysis contributes nearly always to a better understanding of these systems, even though reduction in the strict sense of the word is impossible (chapter 4). Biological systems are open systems; the principles of entropy therefore are not applicable. Owing to their complexity, biological systems are richly endowed with capacities such as reproduction, metabolism, repli-cation, regulation, adaptedness, growth, and hierarchical organization. Nothing of the sort exists in the inanimate world.

Another biology-specific concept is that of *evolution*. To be sure, even before Darwin geologists knew about changes on the Earth's surface and cosmologists were aware of the probability of changes in the universe, particularly in the solar system. However, on the whole, the world was seen as something quite constant, something that had not changed since the day of Creation. This view totally changed after the middle of the nineteenth century when science became aware of the comprehensiveness of the evolution of the living world (chapter 7).

The adoption of the concept of the *biopopulation* is responsible for what now seems probably the most fundamental difference between the inanimate and the living world. The inanimate world consists of Plato's classes, essences, and types, with the members of each class being iden-tical, and with the seeming variation being "accidental" and therefore irrelevant. In a biopopulation, by contrast, every individual is unique, while the statistical mean value of a population is an abstraction. No two of the six billion humans are the same. Populations as a whole do not dif-fer by their essences but only by statistical mean values. The properties of populations change from generation to generation in a gradual man-ner. To think of the living world as a set of forever variable populations grading into each other from generation to generation results in a concept of the world that is totally different from that of a typologist. The Newtonian framework of unalterable laws predisposes the physicist to be a typologist, seemingly almost as if by necessity. Darwin introduced population thinking into biology rather casually, and it took a long time before it was realized that this is an entirely different concept from the typological thinking traditional in the physical sciences (Mayr 1959).

Population thinking and populations are not laws but concepts. It is one of the most fundamental differences between biology and the so-called exact sciences that in biology theories usually are based on concepts while in the physical sciences they are based on natural laws. Examples of concepts that became important bases of theories in various branches of biology are territory, female choice, sexual selection, resource, and geographic isolation. And even though, through appropriate rewording, some of these concepts can be phrased as laws, they are something entirely different from the Newtonian natural laws.

Furthermore, all biological processes differ in one respect fundamentally from all processes in the inanimate world; they are subject to *dual causation*. In contrast to purely physical processes, these biological ones are controlled not only by natural laws but also by *genetic programs*. This duality fully provides a clear demarcation between inanimate and living processes.

The dual causality, however, which is perhaps the most important diagnostic characteristic of biology, is a property of both branches of biology. When I speak of dual causality I am of course not referring to Descartes' distinction of body and soul but rather to the remarkable fact that all living processes obey two causalities. One of them is the natural laws that, together with chance, control completely everything that happens in the world of the exact sciences. The other causality consists of the genetic programs that characterize the living world so uniquely. There is not a single phenomenon or a single process in the living world that is not in part controlled by a genetic program contained in the genome. There is not a single activity of any organism that is not affected by such a program. There is nothing comparable to this in the inanimate world. Dual causation, however, is not the only unique property of biology to support the thesis of the autonomy of biology. Indeed it is reinforced by some six or seven additional concepts. I will now discuss some of these.

The most novel and most important concept introduced by Darwin was perhaps that of *natural selection*. Natural selection is a process that is both so simple and so convincing, that it is almost a puzzle why after 1858 it took almost eighty years before it was universally adopted by evolutionists. To be sure, the process has been somewhat modified in the course of time. It is rather a shock for some biologists to learn that natural selection, taken strictly, is not a selection process at all, but rather a process of elimination and differential reproduction. It is the least adapted individuals that in every generation are eliminated first, while those that are better adapted have a greater chance to survive and reproduce.

There has long been a great deal of argument about what is more important, variation or selection? But there is no argument. The production of variation and true selection are inseparable parts of a single process (chapter 8). At the first step variation is produced by mutation, recombination, and environmental effects, and at the second step the varying phenotypes are sorted by selection. Of course, during sexual selection real selection takes place. Natural selection is the driving force of organic evolution and represents a process quite unknown in inanimate nature. This process enabled Darwin to explain the "design" so important in the arguments of the natural theologians. The fact that all organisms are seemingly so perfectly adapted to each other and to their environment was attributed by the natural theologians to God's perfect design. Darwin, however, showed that it could be equally well, indeed even better, explained by natural selection. This was the decisive refutation of the principle of cosmic teleology (chapter 3).

Evolutionary biology is a historical science

It is very different from the exact sciences in its conceptual framework and methodology. It deals, to a large extent, with unique phenomena, such as the extinction of the dinosaurs, the origin of humans, the origin of evolutionary novelties, the explanation of evolutionary trends and rates, and the explanation of organic diversity. There is no way to explain these phenomena by laws. Evolutionary biology tries to find the answer to "why?" questions. Experiments are usually inappropriate for obtaining answers to evolutionary questions. We cannot experiment about the extinction of the dinosaurs or the origin of mankind. With the experiment unavailable for research in historical biology, a remarkable new heuristic method has been introduced, that of *historical narratives*. Just as in much of theory formation, the scientist starts with a conjecture and thoroughly tests it for its validity, so in evolutionary biology the scientist constructs a historical narrative, which is then tested for its explanatory value.

Let me illustrate this method by applying it to the extinction of the dinosaurs, which occurred at the end of the Cretaceous, about sixty-five million years ago. An early explanatory narrative suggested that they had become the victims of a particularly virulent epidemic against which they had been unable to acquire immunity. However, a number of serious objections were raised against this scenario, which was therefore replaced by a new proposal, according to which the extinction had been caused by a climatic catastrophe. However, neither climatologists nor geologists were able to find any evidence for such a climatic event and this hypothesis also had to be abandoned. However, when the physicist Walter Alvarez postulated that the extinction of the dinosaurs had been caused by the consequences of an asteroid impact on earth, all observations fitted this new scenario. The discovery of the impact crater in Yucatan further strengthened the Alvarez theory. No subsequent observations were in conflict with this theory.

The methodology of historical narratives is clearly a methodology of historical science. Indeed evolutionary biology, as a science, in many respects is more similar to the Geisteswissenschaften than to the exact sciences. When drawing the borderline between the exact sciences and the Geisteswissenschaften, this line would go right through the middle of biology and attach functional biology to the exact sciences while classifying evolutionary biology with the Geisteswissenschaften. This, incidentally, shows the weakness of the old classification of the sciences, which was made by philosophers familiar with the physical sciences and the humanities but ignorant of the existence of biology.

Observation plays as important a role in the physical as in the biological sciences. The experiment is the most frequently used methodology in the physical sciences and in functional biology, while in evolutionary biology the testing of historical narratives and the comparison of a variety of evidence are the most important methods. This methodology is used in the physicalist sciences only in some historical disciplines such as geology and cosmology. The important role of historical narratives in the historical sciences up to now has been almost entirely ignored by philosophers. It is important to point out that comparison is perhaps an even more important and more frequently applied methodology in the biological sciences, from comparative anatomy and comparative physiology to comparative psychology, than is the method of historical narratives. This is also true for molecular biology because comparison is indispensable in most researches in this field. Indeed, much of genomics consists of the comparison of base pair sequences.

Chance

The natural laws usually effect a rather deterministic outcome in the physical sciences. Neither natural nor sexual selection guarantees such determinism. Indeed, the outcome of an evolutionary process is usually the result of an interaction of numerous incidental factors. Chance with respect to functional and adaptive outcome is rampant in the production of variation. During meiosis, in the reduction division it governs both crossing-over and the movement of chromosomes. Curiously, it was this chance aspect of natural selection for which this theory was most often criticized. Some of Darwin's contemporaries, for instance the geologist Adam Sedgwick, declared that invoking chance in any explanation was unscientific. Actually, it is precisely the chanciness of variation that is so characteristic of Darwinian evolution. Even today there is still much argument about the role of chance in the evolutionary process. Selection, of course, always has the last word.

Holistic thinking

Reductionism is the declared philosophy of the physicalists. Reduce everything to the smallest parts, determine the properties of these parts, and you have explained the whole system. However, in a biological system there are so many interactions among the parts – for instance, among the genes of the genotype – that a complete knowledge of the properties of the smallest parts gives necessarily only a partial explanation. Nothing is as characteristic of biological processes as interactions at all levels, among genes of the genotype, between genes and tissues, between cells and other components of the organism, between the organism and its inanimate environment, and between different organisms. It is precisely this interaction of parts that gives nature as a whole, or the ecosystem, or the social group, or the organs of a single organisms, its most pronounced characteristics. As pointed out in chapter 4, rejecting the philosophy of reductionism is not an attack on analysis. No complex system can be understood except through careful analysis. However, the interactions of the components must be considered as much as the properties of the isolated components. How the smaller units are organized into larger units is critically important for the particular properties of the larger units. This aspect of organization and the resulting emergent properties are what the reductionists had neglected.

Limitation to the mesocosmos

As far as their accessibility to the human sense organs is concerned, one can distinguish three worlds. One is the microcosmos or the sub-atomic world of elementary particles and their combinations. The second is the mesocosmos extending from atoms to galaxies, and the third is the macrocosmos, the world of cosmic dimensions. On the whole, only the mesocosmos is relevant to biology, even though in cellular physiology electrons and protons are sometimes involved. To the best of my knowledge, none of the great discoveries made by physics in the twentieth century has contributed anything to an understanding of the living world.

Observation and comparison are highly important methods also in the humanities, and therefore biology functions as an important bridge between the physicalist sciences and the humanities. The foundation of a philosophy of biology is particularly important for the explanation of mind and consciousness. Evolutionary biology has revealed that in such explanations there is no fundamental difference between humans and animals. Evolutionary thinking and the recognition of the role of chance and of uniqueness are now also appreciated in the humanities.

This explains why all earlier endeavors to construct a philosophy of biology within the conceptual framework of the physical sciences were such failures. Biology, we now realize, is indeed largely an autonomous science and a philosophy of biology must be based primarily on the peculiar characteristics of the living world, recognizing at the same time that this is not in conflict with a strictly physicochemical explanation at the cellular-molecular level.[1]

Can an autonomous biology be unified with physics?

In the two hundred years after Galileo there was a unified science; it was physics. There was no biology to cause problems. But the comforting belief in a unified science became increasingly more difficult to uphold with the rise of biology. This difficulty was widely appreciated and whole organizations were founded to undertake a unification of science. The way to accomplish this was through reduction. This view was based on the conviction that all tangible phenomena of this world "are based on material processes that are ultimately reducible . . . to be laws of physics" (Wilson 1998:266). But this suggestion was based on a faulty analysis of biology, neglecting its autonomous components. Such a reduction would be possible only if all of the theories of biology could be reduced to the theories of physics and molecular biology, but this is impossible (see chapter 4). Wilson thought consilience was a mechanism that would make such reduction possible. Indeed he claimed "consilience is the key to unification" (1998:8). And "consilience is to be achieved by reduction to the laws of physics." This is a beautiful dream but none of the autonomous features of biology can possibly be unified with any of the laws of physics. The endeavor of a unification of the sciences is a search for a Fata Morgana. As is said in the vernacular, "you cannot unify apples with oranges."

This conclusion is so important because it has numerous consequences. One of them is that one cannot base a philosophy of biology on the conceptual framework of the physical sciences. Nor can a philosophy of

---

[1] For a review of some of the controversies between supporters and opponents of the autonomy of biology, see Mayr (1996).

biology be expressed by the explanations of a single branch of biology, let us say molecular biology. Instead, it must be based on the facts and fundamental concepts of the entire living world, as was presented in this chapter.

We need a similar analysis of all other sciences and this will permit us to determine what the various sciences have in common. But such analyses, as presented in this chapter for biology, have not yet been undertaken for any of the other sciences.

The importance of biology for the understanding of humans

Until 1859, there was almost complete consensus that humans are fundamentally different from the remainder of creation. Theologians, philosophers, and scientists completely agreed with each other on this point. Darwin's theory of the descent of all species from common ancestors and its application to humans resulted in a fundamental change. One then realized that the human species is a member of the ape family and, is as such, a legitimate object of scientific research. The consequences of this new insight can be seen in the modern developments of anthropology, behavioral biology, cognitive psychology, and sociobiology.

What was perhaps the most shocking finding was how incredibly similar the human genome is to that of the chimpanzee (Diamond 1992). But precisely the comparison with the chimpanzee has led to a better understanding of humans. For instance, it could no longer be denied that many humans have an inborn tendency for strongly aggressive behavior after one discovered that chimpanzees may also show similar aggressive behavior. Yet, altruism also occurs widely among primates (de Waal 1997) and this ancestry facilitates an understanding of human altruism. Comparisons with primates have revealed that it is entirely justified to investigate humans with the same methods used with animals. Part of the philosophy of humans can therefore by merged with biophilosophy.

LITERATURE CITED

Ayala, F. A. 1968. Biology as an autonomous science. *American Scientist*, 56:207–221.

Beatty, J. 1995. The evolutionary contingency thesis. In *Concepts, Theories and Rationality in the Biological Sciences*, G. S. Wolters and J. Lennox (eds.). Pittsburgh: University of Pittsburgh Press, pp. 45–81.

Bergson, H. 1911. *L'Evolution Créatrice*. Paris: Alcan.

Darwin, C. 1859. *On the Origin of Species by Means of Natural Selection or the Preservation of Favored Races in the Struggle for Life*. London: John Murray [1964, Facsimile of the First Edition; Cambridge, MA: Harvard University Press].

Diamond, J. 1991. *The Third Chimpanzee*. New York: Harper Collins.

Driesch, H. 1899. *Philosophie des Organischen*. Leipzig: Quelle und Meyer.

Kitcher, P. 1984. 1953 and all that. *Philosophical Reviews*, 93:335–373.

Mayr, E. 1959. Darwin and the evolutionary theory in biology. In *Evolution and Anthropology. A Centennial*, B. J. Meggers (ed.). Washington, DC: Anthropological Society of America, pp. 1–10.

Mayr, E. 1996. The autonomy of biology: The position of biology among the sciences. *Quarterly Review of Biology*, 71:97–106.

Mayr, E. 2002. Die Autonomie der Biologie [German version]. *Sitzungs Berichte der Gesellschaft Naturforschen der Freunde* (21 Jan. 2002): 5–16.

Rosenberg, A. 1985. *The Structure of Biological Science*. Cambridge: Cambridge University Press.

Ruse, M. 1973. *The Philosophy of Biology*. London: Hutchinson.

Smart, J. J. C. 1963. *Philosophy and Scientific Realism*. London: Routledge & Kegan Paul.

Waal, F. B. M. de. 1997. *Bonobo: The Forgotten Ape*. Berkeley: University of California Press.

Wilson, E. O. 1998. *Consilience*. New York: Alfred A. Knopf.

3

Teleology[1]

Perhaps no other ideology has influenced biology more profoundly than teleological thinking (Mayr 1974, 1988, 1992). In one form or another it was a prevailing world view before Darwin. Appropriately, the discussion of teleology occupies considerable space (10–14%) in several recent philosophies of biology (Beckner 1959, Rosenberg 1985, Ruse 1973, Sattler 1986). This finalistic world view had many roots. It is reflected by the millenarian beliefs of many Christians, by the enthusiasm for progress promoted by the Enlightenment, by transformationist evolutionism, and by everybody's hope for a better future. Such a finalistic world view, however, was only one of several widely adopted Weltanschauungen.

[1] Revised version of Mayr (1992).

39

Three concepts of the world

Grossly simplifying a far more complex picture, one could perhaps distinguish, in the period before Darwin, three ways of looking at the world:

(1) A recently created and constant world. This was the orthodox Christian dogma, which, however, by 1859 had lost much of its credibility, at least among philosophers and scientists (Mayr 1982). This view has been revived in recent years by some fundamentalist Protestant sects.

(2) An eternal and either constant or cycling world, exhibiting no constant direction or goal. Everything in such a world, as asserted by Democritus and his followers, is due to chance or necessity, with chance by far the more important factor. There is no room for teleology in this world view, everything being due to chance or causal mechanisms. It allows for change, but such change is not directional; it is not an evolution. This view gained some support during the scientific revolution and the Enlightenment, but it remained very much a minority view until the nineteenth century. A rather pronounced polarization developed from the seventeenth to the nineteenth centuries, between the strict mechanists, who explained everything purely in terms of movements and forces and who denied any validity whatsoever of the use of teleological language, and their opponents – deists, natural theologians, and vitalists – who all believed in teleology to a lesser or greater extent.

(3) The third view of the world, that of the teleologists, was of a world of long or eternal duration but with a tendency toward improvement or perfection. Such a view existed in many religions, was widespread in the beliefs of primitive people (e.g., the Valhalla of the old Germans), and was represented in Christianity by ideas of a millennium or resurrection. During the rise of deism, after

the scientific revolution and during the era of Enlightenment, there was a widespread belief in the development of ever-greater perfection in the world through the exercise of God's laws. There was a trust in an intrinsic tendency of Nature toward progress or an ultimate goal. Such beliefs were shared even by those who did not believe in the hand of God but who nevertheless believed in a progressive tendency of the world toward ever-greater perfection (Mayr 1988:234–236). It is the belief in cosmic teleology.

Although Christianity was its major source of support, teleological thinking gained increasing strength also in philosophy, from its beginning with the Greeks and Cicero up to the eighteenth and nineteenth centuries. The concept of the *Scala Naturae*, the scale of perfection (Lovejoy 1936), reflected a belief in upward or forward progression in the arrangement of natural objects. Few were the philosophers who did not express a belief in progress and improvement. It also fitted quite well with Lamarck's transformationist theory of evolution, and it is probably correct to say that most Lamarckians were also cosmic teleologists. The concept of progress was particularly strong in the philosophies of Leibniz, Herder, their followers, and, of course, among the French philosophers of the Enlightenment.

What struck T. H. Huxley "most forcibly on his first perusal of the *Origin of Species* was the conviction that teleology, as commonly understood, had received its deathblow at Mr. Darwin's hands" (Huxley 1870:330). However, Huxley's prophecy did not come true. Perhaps the most popular among the non-Darwinian evolutionary theories was that of orthogenesis (Bowler 1983:141–181, 1987), which postulated that evolutionary trends, even nonadaptive ones, were due to an intrinsic drive. Even though the arguments of the orthogenesists were effectively refuted by Weismann (Mayr 1988:499), orthogenesis continued to be highly popular not only in Germany but also in France (Bergson 1911), the United States (Osborn 1934:193–235), and Russia (Berg 1926). The reason was that even though Darwin's demonstration of the nonconstancy of species and of the common descent of all organisms made the acceptance of evolution inevitable, natural selection, the mechanism proposed by Darwin, was so unpalatable to his opponents among the evolutionists that they grasped at any other conceivable mechanism as an anti-Darwinian strategy. One of these was orthogenesis, a strictly finalistic principle (Mayr 1988:234–236) that did not really collapse until the evolutionary synthesis of the 1940s. Simpson (1944, 1949), Rensch (1947), and J. Huxley (1942), in particular, showed that perfect orthogenetic series as claimed by the orthogenesists simply did not exist when the fossil record was studied more carefully; that allometric growth could explain certain seemingly excessive structures; and, finally, that the assertion of the deleteriousness of certain characters, supposed to be due to some orthogenetic force, was not valid. These authors showed, furthermore, that there was no genetic mechanism that could account for orthogenesis.

Both friends and opponents of Darwin occasionally classified him as a teleologist. It is true that this is what he was early in his career, but he gave up teleology after he had adopted natural selection as the mechanism of evolutionary change. Whether this was as late as the 1850s, as claimed by some authors, or already in the early 1840s, as indicated by the research of some historians, is unimportant. There is certainly no support for teleology in the Origin of Species, even though, particularly in his later years and in correspondence, Darwin was sometimes careless in his language (Kohn 1989:215–239). I have previously presented a rather full history of the rise and fall of teleology in evolutionary biology, particularly in Darwin's writings (Mayr 1988:235–255).

All endeavors to find evidence for a mechanism that would explain a general finality in nature were unsuccessful or, where it occurs in organisms, it was explained strictly causally (see below). As a result, by the time of the evolutionary synthesis of the 1940s, no competent biologist was left who still believed in any final causation of evolution or of the world as a whole.

Final causes, however, are far more plausible and pleasing to a layperson than the seemingly so haphazard and opportunistic process of natural selection. For this reason, a belief in final causes had a far greater hold outside of biology than within. Almost all philosophers, for instance, who wrote on evolutionary change in the one hundred years after 1859 were confirmed finalists. All three philosophers closest to Darwin – Whewell, Herschel, and Mill – believed in final causes (Hull 1973). The German philosopher E. von Hartmann (1872) was a strong defender of finalism, stimulating Weismann to a spirited reply. In France, Bergson (1911) postulated a metaphysical force, *élan vital*, which, even though Bergson disclaimed its finalistic nature, could not have been anything else, considering its effects. There is still room for a good history of finalism in the post-Darwinian philosophy, although Collingwood (1945) has made a beginning. Whitehead, Polanyi, and many lesser philosophers were also finalistic (Mayr 1988:247–248).

Refutation of a finalistic interpretation of evolution or of nature as a whole, however, did not eliminate teleology as a problem of philosophy. For the Cartesians, any invoking of teleological processes was utterly unthinkable. Coming from mathematics and physics, they had nothing in their conceptual repertory that would permit them to distinguish between seemingly end-directed processes in inorganic nature and seemingly goal-directed processes in living nature. They feared, as shown particularly clearly by Nagel (1961, 1977), that making such a distinction would open the door to metaphysical, nonempirical considerations. Because all their arguments were based on the study of inanimate objects, they ignored the common view, derived from Aristotle and strongly confirmed by Kant, that truly goal-directed and seemingly purposive processes occur only in living nature. Hence the (physicalist) philosophers ignored the study of living nature and the findings of the biologists. Instead they used teleology to exercise their logical prowess. Why this was so has been explained by Ruse: "What draws philosophers toward teleology is that one has to know, or at least it is generally thought that one has to know, absolutely no biology at all! . . . philosophers want no empirical factors deflecting them in their neo-Scholastic pursuits." (1981:85–101). The irony of this jibe against his fellow philosophers is that, having said this, Ruse himself promptly ignored the literature on teleology written by biologists and concentrated on reviewing the books of three philosophers known for their neglect of biology. Yet Ruse is not alone. One paper or book after the other dealing with teleology continues to be published in the philosophical literature in which the author attempts to solve the problem of teleology with the sharpest weapons of logic, while utterly ignoring the diversity of the phenomena to which the word teleology has been attached, and of course ignoring the literature in which biologists have pointed this out.

Some of the difficulties of the philosophers are due to their misinterpretation of the writings of the great philosophers of the past. Aristotle, for instance, has often been recorded as a finalist, and cosmic teleology has been called an Aristotelian view. Grene is entirely correct when pointing out that Aristotle's *telos* has nothing to do with purpose "either Man's or God's. It was the Judaeo-Christian God who (with the help of neo-Platonism) imposed the dominance of cosmic teleology upon Aristotelian nature. Such sweeping purpose is the very opposite of Aristotelian [philosophy]" (Grene 1972:395–424). Modern Aristotle specialists (Balme, Gotthelf, Lennox, and Nussbaum) are unanimous in showing that Aristotle's seeming teleology deals with problems of ontogeny and adaptation in living organisms, where his views are remarkably modern (Mayr 1988:55–60). Kant was a strict mechanist as far as the inanimate universe is concerned but provisionally adopted teleology for certain phenomena of living nature, which (in the 1790s) were inexplicable owing to the primitive condition of contemporary biology (Mayr 1991:123–139). It would be absurd, however, to use Kant's tentative comments two hundred years later as evidence for the validity of finalism.

The reasons for the unsatisfactory state of the teleological analyses in the philosophical literature are now evident. Indeed, one can go so far as to say that the treatment of the problems of teleology in this literature shows how not to do the philosophy of science. For at least fifty years a considerable number of philosophers of science have written on teleology, basing their analysis on the methods of logic and physicalism, "known to be the best" or at least the only reliable methods for such analyses. These philosophers have ignored the findings of the biologists, even though teleology concerns mostly or entirely the world of life.

They ignored that the word *function* refers to two very different sets of phenomena and that the concept of *program* gives a new complexion to the problem of goal-directedness; they confounded the distinction between proximate and evolutionary causations and between static (adapted) systems and goal-directed activities. Even though there is an enormous philosophical literature on the problems of teleology, those recent books and papers that still treat teleology as a unitary phenomenon are quite useless. No author who had not recognized the differences between the significance of cosmic teleology, adaptedness, programmed goal-directedness, and deterministic natural laws has made any worthwhile contribution to the solution of the problems of teleology.

The principal endeavor of the traditional philosopher was to eliminate teleological language from all descriptions and analyses. They objected to sentences such as "the turtle swims to the shore in order to lay her eggs," or "the wood thrush migrates to warmer climates in order to escape the winter." To be sure, questions that begin with "what?" and "how?" are sufficient for explanation in the physical sciences. However, since 1859 no explanation in the biological sciences has been complete until a third kind of question was asked and answered: "why?" It is the evolutionary causation and its explanation that is asked for in this question. Anyone who eliminates evolutionary "why" questions closes the door on a large area of biological research. Therefore, it is important

45

for the evolutionary biologist to demonstrate that "why?" questions do not introduce a new metaphysical element into the analysis and that there is no conflict between causal and teleological analysis, provided it is precisely specified what is meant by "teleological." I have elsewhere (Mayr 1988:38–66) presented a detailed analysis of the "multiple meanings of teleological" but must present here at least the gist of my findings. Nagel (1977) and Engels (1982) have criticized some of my views. Engels's monograph is the most complete treatment of the teleology problem in the German language. In the following account I have included an answer to the objections of these authors. Before doing so, I want first to clear up a number of assumptions that have been a confusing element in the recent literature. This will allow me to show that the following assertions are invalid.

(1) *Teleological statements and explanations imply the endorsement of unverifiable theological or metaphysical doctrines in science.* This criticism was indeed valid in former times, particularly in the eighteenth and early nineteenth centuries, as well as for most vitalists, including Bergson and Driesch, right up to modern times. It does not apply to any Darwinian who uses teleological language (see below).

(2) *Any biological explanation that is not equally applicable to inanimate nature constitutes rejection of a physicochemical explanation.* This is an invalid objection, because every modern biologist accepts physicochemical explanations at the cellular-molecular level; furthermore, because, as will be shown below, seemingly teleological processes in living organisms can be explained strictly materialistically.

(3) *Teleonomic processes are in conflict with causality because future goals cannot direct current events.* This objection, frequently raised by physicalists, is due to their failure to apply the concept of program, a concept not existing in the classical framework of physicalist concepts and theories.

(4) *Teleological explanations must qualify as laws.* Actually the attempt to insert laws into teleological explanations has led only to confusion (Hull 1982:298–316).

(5) Telos *means either end point or goal; they are the same.* By contrast, for the evolutionary biologist there is a great difference between *telos* as goal and *telos* as endpoint. If one asks whether natural selection and, more broadly, all processes in evolution have a telos, one must be clear which telos one has in mind.

The word *telos* has been used in the philosophical literature with two very different meanings. Aristotle uses it to refer to a process that has a very definite goal, a goal ordinarily anticipated when the process is initiated. The *telos* of the fertilized egg is the adult into which it develops. For the deistic teleologist, cosmic teleology also had a definite goal – i.e., the world in its final perfection as conceived by its creator and effected by His laws. But *telos* has also been used simply to refer to the termination of an end-directed process. The *telos* of a rainstorm is when it stops raining. Day is the *telos* of the night. All processes caused by natural laws sooner or later have an endpoint, but it is misleading to use for this termination the same word *telos* that is ordinarily used for the goal of a goal-directed process. The endpoint of a nonteleological process is, so to speak, an *a posteriori* phenomenon. Pierce (1958, vol. VII, p. 471) realized that the term "teleological" is too strong a word to apply to natural processes in the inorganic world. He therefore suggested that "we might invent the term finious to express their tendency toward a final state."

Many philosophers of science have thought the problem of teleology could be solved by explaining goal-directedness in terms of function – i.e., by translating teleological statements (Wimsatt 1972:1–80) into function statements (Cummins 1975:741–765). Such a translation is also implicit in Hempel (1965), Nagel (1961), and numerous authors since. Whether they recognize six meanings of the term function, as does

Nagel, or ten, as does Wimsatt, all these proposals suffer from the fatal flaw not to have recognized that the word function is used in biology in two very different meanings, which must be carefully distinguished in any teleological analysis. Bock and von Wahlert (1969:269–299) have admirably clarified the situation by showing that function is sometimes used for physiological processes and sometimes for the biological role of a feature in the life cycle of the organism. "For example, the legs of a rabbit have the function of locomotion . . . but the biological role of this faculty may be to escape from a predator, to move toward a source of food, to move to a favorable habitat, [or] to move about in search of a mate." Descriptions of the physiological functioning of an organ or other biological feature are not teleological. Indeed, in favorable cases, they can be largely translated into physicochemical explanations; they are due to proximate causations. What is involved in an analysis of teleological aspects is the biological role of a structure or activity. Such roles are due to evolutionary causations. For this reason in my account I carefully avoid the word function when my concern is the biological role of a feature or process (see below, p. 58).

Categories of teleology

Most philosophers have treated teleology as a unitary phenomenon. This ignores the fact that the term teleological has been applied to several fundamentally different natural phenomena. Under this circumstance, it is no surprise that the search for a unitary explanation of teleology has so far been entirely futile. Beckner (1959) thinks he can distinguish three kinds of teleology, characterized by the terms function, goal, and intention. Although this proposal leads to some ordering of the phenomena, it does not represent a successful solution, owing to the ambiguity of the term function. Woodger (1929) also saw the diverse meanings of the word teleological and attempted to recognize some categories but did not carry the analysis very far. A careful study of all the uses of the term teleological in the philosophical and biological literature leads me to propose a fivefold division. One of the major features of my proposal is to divide the category of function into genuine functional activities and to add the category of adaptedness, corresponding to the history of features with a biological role (see Bock and von Wahlert 1969). Accordingly, I distinguish five different processes or phenomena for which the word teleological has been used:

(1) Teleomatic processes
(2) Teleonomic processes
(3) Purposive behavior
(4) Adapted features; and
(5) Cosmic teleology.

Each of these five processes or phenomena is fundamentally different from the other four and requires an entirely different explanation. The attempts of certain philosophers (most of them!) to find a *unitary explanation* of teleology therefore were totally misconceived. The scientific study of all natural phenomena formerly designated as teleological has deprived the subject teleology of its former mystery. It is now realized that four of the five phenomena traditionally called teleological can be completely explained by science, while the fifth one, cosmic teleology, does not exist. This clarification of the concept of teleology has greatly contributed to the conclusion that biology is a genuine science without any occult properties.

Teleomatic processes

Several philosophers have designated as teleological any processes that "persist toward an end point under varying conditions" or in which "the end state of the process is determined by its properties at the beginning" (Waddington 1957). These definitions would include all processes in inorganic nature that have an endpoint. A river would have to be called teleological because it flows into the ocean. To place such processes in the same category as genuine goal-directed processes in organisms is most misleading.

All objects of the physical world are endowed with the capacity to change their state, and these changes strictly obey natural laws. They are end-directed only in an automatic way, regulated by external forces or conditions – that is, by natural laws. I designated such processes *teleo-matic* (Mayr 1974) to indicate that they are automatically achieved. All teleomatic processes come to an end when the potential is used up (as in the cooling of a heated piece of iron) or when the process is stopped by encountering an external impediment (as when a falling object hits the ground). The law of gravity and the second law of thermodynamics are among the natural laws that most frequently govern teleomatic processes.

Aristotle clearly distinguished teleomatic processes from the teleological ones encountered in organisms and referred to the former as caused "by necessity" (Gotthelf 1976). These are most of the processes called finious by Pierce (1958). They may have an endpoint but they never have a goal. The question "what for?" (*wozu?*) is inappropriate for them. One cannot ask for what purpose lightning had struck a particular tree or for what purpose a flood or an earthquake had killed thousands of people.

Radioactive decay is a teleomatic process; it is not controlled by a program. Any chunk of uranium will experience radioactive decay governed by the same physical laws as any other, in contrast to programs that are highly specific and often unique. The natural laws interact with the intrinsic properties of the material on which they act. Different materials have different properties, and the rate of cooling may differ from one substance to the next. But inherent properties that are the same for any sample of the same substance are something entirely different from a coded program. This is true right down to the molecular level. A given macromolecule has inherent properties, but this by itself is not a program. Programs are formed by a combination of molecules and other organic components.

Prediction is not the defining criterion of a program. If I release a stone from my hand, I can predict that it will fall to the ground. Therefore, says Engels (1982), it is programmed to fall to the ground, and there is no difference between teleomatic and teleonomic processes. This is the same argument Nagel (1977) made with reference to radioactive decay. An example will show how misleading this argument is: somewhere in the mountains a falling stone kills a person. Engels would have to say that this stone was "programmed" to kill a person. The very general terminal situations effected by natural laws are something entirely different from the highly specific goals coded in programs. The existence of programs, of course, is in no way in conflict with natural laws. All the physicochemical processes that take place during the translation and execution of a program strictly obey natural laws. But to neglect the role of information and instruction inevitably results in a most misleading description of a program. Could one explain a computer strictly in terms of natural laws, carefully avoiding any reference to information and instruction?

Teleonomic processes

The term teleonomic has been used with various meanings. When Pittendrigh (1958) introduced the term, he failed to provide it with a rigorous definition. As a result, various authors used it either for programmed functions or for adaptedness as did, for instance, Davis (1961), Simpson (1958), Monod (1970), and Curio (1973). I restricted the term teleonomic to programmed activities (Mayr 1974) and now provide the following definition: *a teleonomic process or behavior is one that owes its goal-directedness to the influence of an evolved program.* The term teleonomic thus implies goal direction of a process or activity. It deals strictly with ultimate causations. They occur in cellular-developmental processes and are most conspicuous in the behavior of organisms. "Goal-directed...behavior is extremely widespread in the organic world; for instance, most activities connected with migration, food-getting, courtship, ontogeny and all phases of reproduction are characterized by such goal orientation. The occurrence of goal-directed processes is perhaps the most characteristic feature of the world of living organisms" (Mayr 1988:45). It is sometimes stated that Pittendrigh and I introduced the term teleonomic as a substitute for the term teleological. This is not correct; rather it is a term for only one of the five different meanings of the highly heterogeneous term teleological.

In my original proposal (Mayr 1974), I suggested that one might expand the application of the term teleonomic to include also the functioning of human artifacts (e.g., loaded dice) that are fixed in such a way as to ensure a wanted goal. This extended use of the term has been criticized, and I now consider that human artifacts are only analogs. Truly teleonomic activities depend on the possession of a genetic program.

All teleonomic behavior is characterized by two components. It is guided by "a program" and it depends on the existence of some endpoint, goal, or terminus that is "foreseen" in the program that regulates the behavior or process. This endpoint might be a structure (in development), a physiological function, the attainment of a geographic position (in migration), or a "consummatory act" (Craig 1916) in behavior. Each particular program is the result of natural selection, constantly adjusted by the selective value of the achieved endpoint.

The key word in the definition of teleonomic is the genetic *program*. The importance of the recognition of the existence of programs lies in the fact that a program is (A) something material and (B) something existing before the initiation of the teleonomic process. This shows that there is no conflict between teleonomy and causality. The existence of teleonomic processes regulated by evolved programs is the reason for the dual causations in biology, due to the natural laws (as in the physical sciences) and due to genetic programs (not found in the physical sciences).

A program might be defined as *coded or prearranged information that controls a process (or behavior) leading it toward a goal*. The program contains not only the blueprint of the goal *but also the instructions for how to use the information of the blueprint*. A program is not a description of a given situation but a set of instructions.

Accepting the concept of program seems to cause no difficulties to a biologist familiar with genetics or to any scientist familiar with the working of computers. However, programs, such as those that control teleonomic processes, do not exist in inanimate nature. Traditional philosophers of science, familiar with only logic and physics, therefore have had great difficulty in understanding the nature of programs, as is well illustrated by the writings of Nagel (1961).

References to the presumed existence of something like a program in the genome or the cells of organisms can be found in the biological literature far back into the nineteenth century. E. B. Wilson, after describing the remarkably teleonomic manner in which the cleavage of an egg takes place, continues: "such a conclusion need involve no mystical doctrine of teleology or of final causes. It means only that the factors by which cleavage is determined are in greater or in less degree bound up with an underlying organization of the egg that precedes cleavage and is responsible for the general morphogenic process. The nature of this organization is almost unknown, but we can proceed with its investigation only on the mechanistic assumption that it involves some kind of material configuration in the substance of the egg" (Wilson 1925). It is important once more to emphasize, because this is almost consistently misunderstood in the classical literature on teleology, that the goal of a teleonomic activity does not lie in the future but is coded in the program. Not enough is known about the genetic-molecular basis of such programs to permit us to say much more than that they are innate or partly innate. The existence of the program is inferred from its manifestations in the behavior of the activities of the bearer of the program. The existence of these genetic programs in organisms (= ultimate causes) is the result of proximal causes having acted during the past evolutionary history of the organisms.

Concepts, corresponding to a program, go back all the way to antiquity. After all, Aristotle's *eidos* had many of the properties we now ascribe to the genetic program, as was pointed out by Jacob (1970) and Delbrück (1971). So did Buffon's *moule intérieur* (Roger 1989) as well as the many speculations about inborn memories from Leibniz and Maupertuis to Darwin, Hering, and Semon. As sound as the intuition of these thinkers had been, it required an understanding of the DNA nature of the genome before the genetic program could be considered a valid scientific concept.

The study of teleonomic programs has shown that several kinds can be distinguished. A program in which complete instructions are laid down in the DNA of the genotype is called a *closed program* (Mayr 1964). Most programs that control the instinctive behavior of insects and of lower invertebrates seem to be largely closed programs. It is not yet known to what extent new information can be incorporated into supposedly closed programs. There are, however, other types of programs, *open programs*, which are constituted in such a way that additional information can be incorporated during a lifetime, acquired through learning, conditioning, or other experiences. Most behavior in higher animals is controlled by such open programs. Their existence has long been known to ethologists without their introducing a special terminology. In the famous case of the following reaction of the young gosling, the open program provides for the "following reaction," but the particular object (the "parent") to be followed is added by experience (by "imprinting"). Open programs are very frequent in the behavior program of higher organisms, but even in some invertebrates there is often opportunity to make use of individual experience in filling out open programs – for instance, with respect to suitable food or potential enemies or the nest site in solitary wasps.

The programs controlling teleonomic activities initially were thought of exclusively in terms of the DNA of the genome. However, in addition to such genetic programs it might be useful to recognize *somatic programs*. "For instance, when a turkey gobbler displays to a hen, his display movements are not directly controlled by the DNA in his cell nuclei, but rather by a somatic program in his central nervous system. To be sure, this neuronal program was laid down during development under the partial control of instructions from the genetic program. But it is now an independent somatic program" (Mayr 1988:64). Somatic programs are particularly important in development. Each stage in ontogeny, together with relevant environmental circumstances, represents, so to speak, a somatic program for the next step in development. Most of the embryonic structures that have been cited as evidence for recapitulation, like the gill arches of tetrapod embryos, are presumably somatic programs. The reason why they have not been removed by natural selection is that this would have seriously interfered with subsequent development. The existence and role of somatic programs have been understood by embryologists at least since Kleinenberg (1886).

The various kinds of programs are not separated by sharp boundaries. All are the result of proximal causes that have acted during the past evolutionary history of the organism. And all are associated with the concept of ultimate causations.

To borrow the term program from informatics is not a case of anthropomorphism. There is a great deal of equivalence between the "program" of the information theorists and the genetic and somatic programs of the biologists. The origin of a program is quite irrelevant for its definition. It can be the product of evolution, as are all genetic programs, or it can be the acquired information of an open program. It can be genetic, whether closed or open, or it can be somatic when additional information, acquired during the life of the individual, is added to the instruction of the genetic program.

An objection that has been raised against the concept of program is that reflexes would then also be teleonomic activities. Why not? Some of them undoubtedly are. Sherrington (1906:235) was fully aware of the significance of the reflex as an adapted act. He said "the purpose of a reflex seems as legitimate and urgent an object for natural enquiry as the purpose of the coloring of an insect or blossom. And the importance to physiology is, that the reflex cannot be really intelligible to the physiologist until he knows its aim." The eyelid clearly is programmed to close by reflex when a threatening object or disturbance approaches the eye. A similar adaptive function is evident for numerous reflexes. Other reflexes, like the knee-jerk reflex so beloved by physicians, seem to be merely an irrelevant property of certain nerves, as irrelevant as the heart sounds are for the functioning of the heart. It would be most useful if a neurophysiologist would someday analyze the better-known reflexes for any possible adaptive significance.

The directedness of a teleonomic action is effected by a number of devices – first of all, of course, by the program itself; but the program does not induce a simple unfolding of some completely preformed gestalt, for it always controls a more or less complex process that must allow for internal and external disturbances. Teleonomic processes during ontogenetic development, for instance, are constantly in danger of being derailed even if only temporarily. Waddington (1957) has quite rightly called attention to the frequency and importance of homeostatic devices that correct such deviations; they virtually guarantee the appropriate canalization of development.

Negative feedbacks play an important role not only in development but also in many other teleonomic processes. They are, however, not the essence of the teleonomic activity. As I pointed out earlier, "the truly characteristic aspect of goal-seeking behavior is not that mechanisms exist which improve the precision with which a goal is reached, but rather that mechanisms exist which initiate, i.e., 'cause' this goal-seeking behavior" (Mayr 1988:46).

Purposive behavior in thinking organisms

Several philosophers have used human intentions and purposive acts as exemplary illustrations of teleological processes. This introduces concepts such as purpose, intention, and consciousness into the discussion and makes it an aspect of human psychology. But this field is highly controversial and, in an earlier treatment of teleology (Mayr 1992), I therefore excluded purposive behavior from the discussions.

Much recent work in animal behavior has since convinced me that I was mistaken. Purposive behavior that is clearly goal-directed is widespread among animals, particularly among mammals and birds, and fully qualifies to be called teleological. Several species of jays bury in the fall acorns and piñon seeds and return to these caches (which they remember remarkably accurately) and retrieve this food, when at the end of the winter natural sources of food are largely exhausted. The literature on animal behavior is full of descriptions of animal behavior that is clearly purposive, revealing careful planning. Another famous example is the hunting strategy of lionesses. When preparing an attack, the pride may separate into two groups, one of which moves behind the intended victim, cutting off its escape route. In such purposive planning, there is in principle no difference between humans and thinking animals.

Adapted features

Features that contribute to the adaptedness of an organism are in the philosophical literature usually referred to as teleological or functional systems. Both of the designations are potentially misleading. These features are stationary systems, and as I pointed out previously (Mayr 1988:51–52), the word teleological would not seem to be appropriate for phenomena that do not involve movements.

The designation teleological system is misleading for a second reason. It was adopted by the older philosophical literature under the assumption that these features had originated through some teleological force of nature. This assumption was largely a heritage of natural theology, with its belief that the usefulness of each feature had been given by God. The fallacy of this thinking has been refuted particularly effectively by Dawkins in his splendid book *The Blind Watchmaker* (1986). Immanuel Kant's interest in teleology focused on adapted features. Owing to the scant knowledge of biology available at the end of the eighteenth century, he was unable to provide a causal explanation. He therefore ascribed adaptedness to teleological forces, by which he presumably meant the hand of God (Mayr 1988:57–59, 1991). Since 1859 such defeatism has become unnecessary. Darwin has taught us that seemingly teleological evolutionary changes and the production of adapted features are simply the result of variational evolution, consisting of the production of large amounts of variation in every generation, and the probabilistic survival of those individuals remaining after the elimination of the least-fit phenotypes. Adaptedness thus is an a posteriori result rather than an a priori goal seeking. For this reason, the word teleological is misleading when applied to adapted features.

Nor should they be called functional systems owing to the confusing dual meaning of the word function. Indeed most of those who use the terminology functional systems were referring to the biological role of these features and their effectiveness in carrying out this role. Proximate and ultimate (evolutionary) causations were frequently confounded in functionalist discussions. Munson (1971) and Brandon (1981) have excellently stated the reasons why an adaptationist language, in connection with adapted features, and in connection with an answer to "what for?" questions, is to be preferred to teleological or functional language.

One of the characteristics of adapted features is that they can perform teleonomic activities. They are, so to speak, executive organs for teleonomic programs. Therefore, I have suggested (Mayr 1988) that they perhaps might be considered to be somatic programs.

More than anything else it is the existence of adapted features that led biologists to ask "why?" questions. The first area in biology where they were used was in physiological research. When Harvey was asked what had induced him to think of the circulation of blood, he answered, "I wondered why there were valves in the veins" (Krafft 1982). Evidently they permit only a one-directional flow of the blood and this, almost automatically, led to an assumption of circulation. One physiological discovery after another resulted from asking "why?" questions concerning organs with unknown function. Such "why?" and "what for?" questions eventually became equally productive in other branches of biology, and the heuristic value of this methodology has by no means been exhausted.

Cosmic teleology

Before the nineteenth century, the belief was almost universal that change in the world was due to an inner force or tendency toward progress and to ever-greater perfection (see above). Gillispie (1951), Glacken (1967), and I (Mayr 1982) have described the immense power of this ideology. As late as 1876 K. E. von Baer made a passionate plea for the recognition of finalism to give pleasure to those people "who consider the world and particularly the organic world as the result of a development which tends toward higher goals and is guided by reason" (1876). The most determined opponents of natural selection were teleologists, and teleological theories of evolution (orthogenesis, etc.) continued to be dominant until the beginning of the twentieth century (Kellogg 1907; Mayr 1982; Bowler 1983, 1987).

When it was being realized that the world was neither recent nor constant, three categories of explanations for seemingly finalistic changes were advanced:

(1) These changes are due to the action of an evolutionary planner (theistic explanation).

(2) These changes are guided by a built-in program, analogous to a teleonomic program in the genotype of an individual (orthogenetic explanations). Much of the post-Darwinian research resulted in providing evidence that such a cosmic program does not exist and that the irregularities of cosmic evolution are far too great to be reconciled with the existence of a program. Indeed by the time of the evolutionary synthesis (1930s to 1940s) all support for orthogenetic theories had disappeared.

(3) There is no cosmic teleology; there is no trend in the world toward progress or perfection. Whatever changes and trends in the cosmos are observed in the course of the world's history, they are the result of the action of natural laws and natural selection. This third explanation fits the observed facts so well that it makes it unnecessary to invoke the other two explanations.

The refutation of cosmic teleology leaves us with one unsolved problem: how can one explain the seemingly upward trend in organic evolution? Author after author has referred to the progression from the lowest prokaryotes (bacteria) to the nucleated eukaryotes, the metazoans, warm-blooded mammals and birds, and finally humans with their elaborate brains, speech, and culture. The defenders of orthogenesis never tired of claiming that this was irrefutable evidence for some intrinsic power in living nature toward progress, if not even to an ultimate goal. Again, it was Darwin who showed that such an assumption was not inevitable. The process of natural selection, acting in every population, generation for generation, is indeed a mechanism that favors the rise of ever better-adapted species; it favors the invasion of new niches and adaptive zones; and as the end-result of competition among species it would favor development of what are best described as advanced types. Descriptively there is no question about what has happened during the diverse steps from the most primitive bacteria to humans. Whether one is justified in referring to this as progress is still controversial. This much is clear, however; natural selection provides a satisfactory explanation for the course of organic evolution and makes an invoking of supernatural teleological forces unnecessary. And those who accept the occurrence of advance or progress in evolution do not ascribe it as due to teleological forces or tendencies but rather as the product of natural selection.

The current status of teleology

The removal of the mentioned four material processes from the formerly so heterogeneous category "teleological" leaves no residue. This proves the nonexistence of cosmic teleology.

The recognition that four seemingly teleological processes – that is, teleonomic processes, teleomatic processes, the achievement of adaptedness by natural selection, and purposive behavior – are strictly material phenomena has deprived teleology of its former mystery and supernatural overtones. There is adaptedness (Kant's *Zweckmässigkeit*) in living nature but Darwin showed that its origin can be explained materialistically. Even though there are indeed many organic processes and activities that are clearly goal-directed, there is no need to involve supernatural forces, because the goal is already coded in the program that directs these activities. Such teleonomic processes, in principle, can be reduced to chemicophysical causes. Finally, there are all the end-achieving processes in inorganic nature that are simply due to the operation of natural laws such as gravity and the laws of thermodynamics. None of the four recognized teleological processes works backward from an unknown future goal; there is no backward causation. This refutes the formerly frequently made claim of a conflict between causal and teleological explanations. Such a claim might be true if cosmic teleology existed, but it is invalid for the four kinds of teleology now accepted by science.

Teleology and evolution

After Darwin established the principle of natural selection, this process was widely interpreted to be teleological, both by supporters and by opponents. Evolution itself was frequently considered a teleological process because it would lead to "improvement" or "progress" (Ayala 1970). Perhaps such an interpretation was not altogether unreasonable in the framework of the Lamarckian transformational paradigm. However, it is no longer a reasonable view when one fully appreciates the variational nature of Darwinian evolution, which has no ultimate goal and which, so to speak, starts anew in every generation. At best the process of natural selection may fit the definition of Pierce's "finious" processes (Pierce 1958, Short 1984); yet considering how often natural selection leads into fatal dead ends and considering how often during evolution its premium changes, resulting in an irregular zigzag movement of the evolutionary change, it would seem singularly inappropriate to use the designation teleological for any form of directional evolution. To be sure, natural selection is an optimization process, but it has no definite goal, and, considering the number of constraints and the frequency of chance events, it would be most misleading to call it teleological. Nor is any improvement in adaptation a teleological process, because whether a given evolutionary change qualifies as a contribution to adaptedness is strictly a post hoc decision. None of the fifteen authors contributing to a recent volume on natural selection and optimization during evolution (Dupré 1987) has used the term teleological.

This has to be remembered when one encounters teleological language in evolutionary interpretations (O'Grady 1984). When an author says that species have evolved isolating mechanisms to protect their genetic integrity, it simply means that individuals avoiding hybridization with individuals of other species had greater reproductive success than those that hybridized. Therefore, a genetic predisposition not to hybridize was rewarded with reproductive success (Mayr 1988). Natural selection deals with properties of individuals of a given generation; it simply does not have any long-range goal, even though this may seem so when one looks backward over a long series of generations. Alas, some authors even in the most recent literature seem to endow evolution with a teleological capacity. As recently as 1985 J. H. Campbell said "It becomes increasingly evident that organisms evolve special structures to promote their capacities to evolve, and that these structures enormously expand the scope of the evolutionary process. Nevertheless, function is fundamentally a teleological concept, especially when applied to the evolutionary process" (Campbell, 1985). As Munson (1971) has rightly pointed out, such a dubious use of the word teleological can easily be avoided by using adaptationist language.

LITERATURE CITED

Ayala, F. J. 1970. Teleological explanation in evolutionary biology. *Philosophy of Science*, 37:1–15.

Baer, K. E. von. 1876. *Studien aus der Geschichte der Naturwissenschaften*. St. Petersburg: H. Schmitzdorf.

Beckner, M. 1959. *The Biological Way of Thought*. New York: Columbia University Press.

Berg, L. S. 1926. *Nomogenesis, or Evolution determined by Law*. London: Constable.

Bergson, H. 1911. *Creative Evolution*. Paris: Alcan.

Bock, W. J., and G. von Wahlert. 1969 Adaptation and the form-function complex. *Evolution*, 19:269–299.

Bowler, P. J. 1983. *The Eclipse of Darwinism*. Baltimore: John Hopkins University Press.

Bowler, P. J. 1987. *The Non-Darwinian Revolution*. Baltimore: John Hopkins University Press.

Brandon, R. N. 1981. Biological teleology: Questions and explanation. *Studies in the History and Philosophy of Science*, 12:91–105.

Campbell, J. 1985. An organizational interpretation of evolution. In *Evolution at a Crossroads*, D. Depew and B. H. Weber (eds.). Cambridge, MA: MIT Press, pp. 133–167.

Collingwood, R. G. 1945. *The Idea of Nature*. Oxford: Clarendon Press.

Craig, W. 1916. Appetites and aversions as constituents of instincts. *Biological Bulletin*, 34:91–107.

Cummins, R. 1975. Functional analysis. *Journal of Philosophy*, 72:741–765.

Curio, E. 1973. Towards a methodology of teleonomy. *Experientia*, 29:1045–1058.

Davis, B. D. 1961. The teleonomic significance of biosynthetic control mechanisms. *Cold Spring Harbor Symposia*, 26:1–10.

Dawkins, R. 1986. *The Blind Watchmaker*. London: W. W. Norton.

Delbrück, M. 1971. Aristotle-totle-totle. In *Of Microbes and Life*, J. Monod and E. Borek (eds.). New York: Columbia University Press, pp. 50–55.

Dupré, J., ed. 1987. *The Latest on the Best: Essays on Evolution and Optimality*. Cambridge, MA: MIT Press.

Engels, E.-M. 1982. *Die Teleologie des Lebendigen*. Berlin: Duncker & Humblot.

Gillispie, C. C. 1951. *Genesis and Geology*. New York: Harper & Bros.

Glacken, C. J. 1967. *Traces on the Rhodian Shore. Nature and Culture in Western Thought from Ancient Times to the End of the Eighteenth Century*. Berkeley: University of California Press.

Gotthelf, A. 1976. Aristotle's conception of final causality. *Review of Metaphysics*, 30:226–254.

Goudge, T. A. 1961. *The Ascent of Life*. Toronto: University of Toronto Press.

Grene, M. 1972. Aristotle and modern biology. *Journal of Historical Ideas*, 33:395–424.

Hartmann, E. v. 1872. *Das Unbewusste vom Standpunkt der Physiologie und Deszendenzlehre*. Berlin: C. Duncker.

Hempel, C. G. 1965. *Aspects of Scientific Explanation*. New York: Free Press.

Hull, D. L. 1973. *Darwin and His Critics*. Cambridge, MA: Harvard University Press.

Hull, D. L. 1982. Philosophy and biology. *Contemporary Philosophy*, 2:298–316.

Huxley, J. 1942. *Evolution: The Modern Synthesis*. London: Allen & Unwin.

Huxley, T. H. 1870. *Lay Sermons, Addresses and Reviews*. London.

Jacob, F. 1970. *La Logique du Vivant*. Paris: Gallimard.

Kellogg, V. L. 1907. *Darwinism Today*. New York: Henry Holt.

Kleinenberg, N. 1886. Über die Entwicklung durch Substitution von Organen. *Zeitschrift für wissenschaftliche Zoölogie*, pp. 212–224.

Kohn, D. 1989. Darwin's ambiguity: The secularization of biological meaning. *British Journal for the History of Science*, 22:215–239.

Kolb, D. 1992. Kant, teleology, and evolution. *Synthese*, 91:9–28.

Krafft, F. 1982. Die Idee der Zweckmässigkeit in der Geschichte der Wissenschaften. *Berichte zur Wissenschaftsgeschichte*, 5:1–152.

Kullmann, W. 1979. *Die Teleologie in der aristotelischen Biologie*. Heidelberg: C. Winter.

Lenoir, T. 1982a. *The Strategy of Life*. Dordrecht: D. Reidel.

Lenoir, T. 1982b. Teleology without regrets. *Studies in the History and Philosophy of Science*, 12:293–353.

Lovejoy, A. O. 1936. *The Great Chain of Being*. Cambridge, MA: Harvard University Press.

Mayr, E. 1964. The evolution of living systems. *Proceedings of the National Academy of Sciences*, 51:934–941.

Mayr, E. 1974. Teleological and teleonomic. A new analysis. *Boston Studies in the Philosophy of Science*, 14:91–117.

Mayr, E. 1982. *The Growth of Biological Thought*. Cambridge, MA: Harvard University Press.

Mayr, E. 1988. *Toward a New Philosophy of Biology*. Cambridge, MA: Harvard University Press.

Mayr, E. 1991. The ideological resistance to Darwin's theory of natural selection. *Proceedings of the American Philosophical Society*, 135:123–139.

Mayr, E. 1992. The idea of teleology. *Journal of Historical Ideas*, 53:117–135.

Mayr, E. 1998. The multiple meanings of teleological. *History and Philosophy of the Life Sciences*, 20:35–40.

Monod, J. 1970. *Le Hasard et la Necessité*. Paris: Seuil.

Munson, R. 1971. Biological adaptation. *Philosophy of Science*, 38:200–215.

Nagel, E. 1961. *The Structure of Science*. New York: Harcourt, Brace & World.

Nagel, E. 1977. Teleology revisited: goal directed processes in biology. *Journal of Philosophy*, 74:261–301.

O'Grady, R. T. 1984. Evolutionary theory and teleology. *Journal of Philosophy*, 74:261–301.

Osborn, H. F. 1934. Aristogenesis, the creative principle in the *Origin of Species*. *American Naturalist*, 68:193–235.

Pierce, C. S. 1958. *Collected Papers*, A. W. Burks (ed.). Cambridge, MA: Harvard University Press, Vol. VII, pp. 298–316.

Pittendrigh, C. S. 1958. Adaptation, natural selection and behavior. In *Behavior and Evolution*, A. Roe and G. G. Simpson (eds.). New Haven: Yale University Press, pp. 390–416.

Rensch, B. 1947. *Neuere Probleme der Abstammungslehre*. Stuttgart: Enke.

Roger, J. 1989. *Buffon*. Paris: A. Fayard.

Rosenberg, A. 1985. *The Structure of Biological Science*. Cambridge: Cambridge University Press.

Ruse, M. 1973. *The Philosophy of Biology*. London: Hutchinson.

Ruse, M. 1981. The last word on teleology, or optimality modes vindicated. In *Is Science Sexist?*, M. Ruse (ed.). Cambridge: Cambridge University Press, pp. 85–101.

Sattler, R. 1986. *Biophilosophy*. Berlin: Springer-Verlag.

Sherrington, C. S. 1906. *The Integrative Action of the Nervous System*. New Haven: Yale University Press.

Short, T. L. 1984. Teleology in nature. *American Philosophical Quarterly*, pp. 311–320.

Simpson, G. G. 1944. *Tempo and Mode in Evolution*. New York: Columbia University Press.

Simpson, G. G. 1949. *The Meaning of Evolution*. New Haven: Yale University Press.

Simpson, G. G. 1958. Behavior and evolution. In *Behavior and Evolution*, A. Roe and G. G. Simpson (eds.). New Haven: Yale University Press, pp. 507–535.

Sterelny, K., and P. E. Griffith. 1999. *Sex and Death*. Chicago: University of Chicago Press.

Waddington, C. H. 1957. *The Strategy of the Genes*. London: Allen & Unwin.

Wilson, E. B. 1925. *The Cell in Development and Heredity*, 3rd ed. New York: Macmillan, p. 1005.

Wimsatt, W. 1972. Teleology and the logical status of function statements. *Studies in the History and Philosophy of Science*, 3:1–80.

Woodger, J. H. 1929. *Biological Principles*. London: Routledge & Kegan Paul.

4

Analysis or Reductionism?

IT IS ONLY COMMON SENSE to believe that a complex phenomenon cannot be fully understood unless it is dissected into smaller components, each of which must be studied separately. This approach was already adopted in principle by the Ionian philosophers when they reduced natural phenomena to four basic elements – earth, water, air, and fire – and analysis has been a tradition in philosophy and science ever since. The anatomist did not study the body as a whole but attempted to understand its workings by dissecting it into the component organs, nerves, muscles, and bones. The objective of microscopy was the study of smaller and smaller components of tissues and cells. The endeavor to carry the analysis to ever-lower levels, to ever-smaller components, was at first motivated primarily because it is such a heuristic methodology.

Much of the history of biology is a tale of the triumphs of this analytical approach. Organic diversity as a whole was unmanageable until organisms were segregated into species. The cell theory of Schwann and Schleiden was such a success because it showed that both plants and animals consisted of the same basic structural elements: cells. Physiology made its most important findings through the most careful analysis of the major organs down to cells and macromolecules. And a similar success of analysis can be shown for any biological discipline. Owing to this unbroken history of successes, no one would question the heuristic importance of analysis.

The mechanists, in their opposition to vitalism, demanded that all living phenomena be analyzed down to the lowest component elements to show that no residue was left after everything had been explained in terms of physics and chemistry. This culminated in the famous Berlin declaration of the physiologists Brücke, DuBois Reymond, and Helmholtz, "to promote the truth that no other forces are active in the organic world than the common physico-chemical ones." They limited their claim to forces; they did not apply it to systems, concepts, or even processes. Yet, the explanatory power of this approach seemed so persuasive that even the naturalist Weismann glibly spoke of certain biological processes as being due to the "movement of molecules."

At a later time, when trying to explain biological phenomena in terms of chemistry and physics, one referred to this process no longer as analysis but as *reduction*. This term often was rather misleading as later events showed. The reductionists called their opponents anti-reductionists, again an unfortunate term because most of them were simply nonreductionists who carried their analysis only to that lowest level where it still yielded useful information. They were not reductionists because they did not adopt the belief of the scientific reductionists that "in principle" everything in living nature could be reduced to chemistry and physics (Mayr 1988) or adopt the belief that everything in science could be fully understood at the lowest level of organization.

The composition of the two camps, the reductionists and the nonreductionists, changed rather drastically over time. As long as vitalism was still alive and was promoted by distinguished authors such as Driesch, Bergson, J. S. Haldane, Smuts, and Meyer-Abich, all nonvitalistic biologists more or less adopted the reductionist credo. However, after vitalism had become obsolete, a belief in strict reductionism was more and more confined to the physicalists, while most biologists adopted a holistic organicism. They accepted constructive analysis but rejected the more extreme forms of reductionism.

Until far into the twentieth century philosophers almost consistently confounded analysis and reduction. However, to have isolated all the parts, even the smallest ones, is not enough for a complete explanation of most systems, as claimed by the reductionists. For a complete explanation one also needs to understand the interaction among these parts. As T. H. Huxley pointed out a long time ago, partitioning water into hydrogen gas and oxygen gas does not explain the liquidity of water.

An approach that includes a study of the interactions of higher levels in a complex system is called a holistic approach. It is in conflict with the manifold attempts of philosophers, physicalists, and some biologists "to reduce biology to physics and chemistry."

If the claims of the reductionists were true that any phenomenon requires for its full explanation only a complete dissection into its smallest parts and an explanation of the properties of these smallest parts, then the importance of each branch of science would be the greater the nearer it is to the level of these smallest parts. Needless to say, the workers in the more complex branches of science saw in this claim only a ploy of the chemists and physicists to boost the importance of their fields. As Hilary Putnam said correctly: "What [reductionism] breeds is physics worship coupled with neglect of the 'higher-level' sciences. Infatuation with what is supposedly possible in principle goes with indifference to practice and to the actual structure of *practice*" (1973).

Reductionist rivalry existed not only between sciences but also within them. In the days when molecular biology thought it was about to replace all other branches of biology, biochemist George Wald said that molecular biology is not to be thought of as a special field or a different kind of biology, "it is the whole of biology" (Wald 1963). In the same spirit of reductionist hubris one philosopher patronizingly conceded that "research in classical biology may be of value," [but only "may be"!] (Schaffner 1967). With statements such as these it is not surprising that for a while the argument about reductionism became rather heated.

In retrospect one wonders why the problem of reduction could ever have become such a dominant component of the philosophy of biology, as for instance in the treatments of Ruse (1973) and Rosenberg (1985). As Hull rightly said, "there is certainly more to the philosophy of biology than whether or not biology can be reduced to chemistry and physics" (1969b:251).

What is the crucial difference between the concepts analysis and reduction? The practitioner of analysis claims that the understanding of a complex system is facilitated by dissecting it into smaller parts. Students of the functions of the human body chooses as their first approach its dissection into bones, muscles, nerves, and organs. They make neither of two claims made by the reductionists (A) that the dissection should proceed "down to the smallest parts," – i.e., atoms and elementary particles, and (B) that such a dissection will provide a complete explanation of the complex system. This reveals the nature of the fundamental difference between analysis and reduction. Analysis is continued downward only as long as it yields useful new information and it does not claim that the "smallest parts" give all the answers.

Kinds of reduction

When one reads the literature on reduction, one is astonished and rather dismayed at the heterogeneity of the use of the term reduction. In due time it became very obvious that a classification of the different kinds of reduction was needed, and this indeed was attempted by a number of authors. The subject is referred to in logic and in other branches of philosophy and in various nonbiological branches of science. Best known is the attempted reduction of thermodynamics to mechanics. Popper (1974) has excellently described the limited successes, but mostly failures, of reduction in the physical sciences. In my own account I will leave aside all discussions of reduction not dealing with biology. For a more technical treatment of reductionism, see Hoyningen-Huene (1989).

Analysis

The first step of clarification is to make a clear distinction between analysis and reduction. The method of analysis consists of dissecting a more or less complex system into its components, all the way down to the molecular level if this is productive. This permits the separate study of each component. This is a continuation of the historical approach that led from gross anatomy to microscopy and from organ physiology to cellular physiology. As useful as analysis is, it has severe limitations in its application. In biology it has been applicable, in the strictest sense, only to the study of proximate causations. As both Simpson (1974) and Lewontin (1969) have shown, the physicochemical approach is totally sterile in evolutionary biology. The historical aspects of biological organization are entirely out of reach of physicochemical reductionism.

Analysis differs from reduction by not claiming that the components of a system, revealed by analysis, provide complete information on all the properties of a system, because analysis does not supply a full description of the interactions among the components of a system. In spite of its being a highly heuristic method for the study of complex systems, it would be an error to refer to analysis as reduction.

Explanatory reduction

The proponents of strict reduction make one or both of the following two claims.

(1) No higher-level biological phenomenon can be understood until it has been analyzed into the components of the next lower level; this process is to be continued downward to the level of the purely physicochemical components and processes.

(2) As a consequence of this line of reasoning, it is also claimed that a knowledge of the components at the lowest level permits the reconstruction of all higher levels and provides exhaustively an understanding of these higher levels. These claims of the reductionists are based on their conviction that wholes are not more than the additive sums of their parts – emergent properties do not exist.

Experience has shown that these claims of the reductionists are only rarely validated. Let me list a number of reasons for this failure:

What counts in the study of a complex system is its organization. Descending to a lower level of analysis often decreases the explanatory power of the preceding analysis (Kitcher 1984:348). No one would be able to infer the structure and function of a kidney even if given a complete catalog of all the molecules of which it is composed.

This argument is valid not only for complex biological systems but also for inanimate ones. If I want to understand the nature and function of a hammer, I apply the appropriate laws of mechanics. If I were to try to analyze the hammer at the next lower level, and determine, for instance, of what kind of wood the handle is made, if I would then study the structure of this wood through the microscope, and continue downward through chemistry to the constituent molecules, atoms, and elementary particles of which the handle is composed, I would add absolutely nothing to the understanding of the properties of the hammer *as a hammer*. Indeed, the handle could be made of a plastic (as are some modern hammers) or of a tough light metal. It is the combination of handle (stem) and hammerhead that constitutes the hammer and permits the explanation of its function. A further downward analysis adds nothing.

One could produce thousands of examples that would demonstrate equally convincingly how wrong the claim is that downward analysis of a system to the next lower level of integration automatically leads to a better and more complete understanding. Actually in the course of downward analysis invariably a level is reached sooner or later where the whole meaning of the system is destroyed when the analysis is carried downward any further.

The most down-to-earth physicists confess that the spectacular advances of solid state and of elementary particle physics actually have not had any impact on our concept of the middle world. This is a confession that is rather painful for reductionists, who at one time had so loudly proclaimed that all the remaining mysteries of the world would be solved as soon as we could build even bigger atom smashers. In fact, it is now quite evident that even an exhaustive knowledge of protons, neutrinos, quarks, electrons, and whatever other elementary particles there may be, would not help us one single bit in explaining the origin of life, in explaining differentiation during ontogeny, or in explaining mental activities in the central nervous system. Opposing claims, so often made by overenthusiastic reductionists, are without any foundation.

This does not deny that analysis *occasionally* produces "upward illumination." For instance, the discovery of the structure of DNA by Watson and Crick made it possible to explain two major properties of DNA — its mode of replication and that of information transfer. However, both belong to the same hierarchical level.

The consistent failure of explanatory reductionism indicates that a different approach must be taken in biological analysis, based (A) on the insight that all biological systems are ordered systems, which owe much of their properties to this organization and not simply to the chemical-physical properties of the components; (B) on the insight that there is a system of levels of organization with the properties of the higher systems not necessarily reducible to (or explained by) those of the lower ones; (C) on the recognition that biological systems store historically acquired information, not accessible to a physicalist reductionist analysis; and (D) on the recognition of the frequency of the occurrence of emergence. In complex systems properties often emerge that are not displayed by (and cannot be predicted from) a knowledge of the components of these systems.

Emergence

Emergence, the occurrence of unexpected characteristics in complex systems, has long been a highly controversial subject in the philosophy of biology. Does it really occur, and, if so, what causes it? Is it necessarily an indication of metaphysical or supernatural factors?

As Mandelbaum (1971:380) has pointed out, the view that composite wholes have properties not evident in their components has been widely accepted since the middle of the nineteenth century. The principle was already enunciated by Mill, but it was Lewes (1875) who not only presented a thorough analysis of the topic but also proposed the term *emergence* for this phenomenon. Valuable treatments of the subject are presented by Goudge (1965), Mandelbaum (1971), Ayala and Dobzhansky (1974), and Mayr (1982:63, 863). Lloyd Morgan in his work *Emergent Evolution* (1923) made the concept particularly widely known. For Popper (1974:269) the term indicates "an apparently unforeseeable evolutionary step," and the term emergence was therefore particularly often used in connection with the evolutionary origin of life, mind, and human consciousness.

Their attitude toward emergence is the most decisive difference between reductionists and nonreductionists (= holists). For reductionists, wholes are not more than the additive sums of their parts; they do not possess emergent properties. For the holist the properties and modes of action at a higher level of integration are not exhaustively explicable by the summation of the properties and modes of action of their components taken in isolation. This thought is well expressed in the classical statement that "the whole is more than the sum of its parts." Believing that the term emergence implies something metaphysical, a number of other terms have been introduced for this phenomenon, such as *fulguration* by Lorenz (1973) and *compositionism* by Simpson (1964) and Dobzhansky (1968).

During its long history, the term "emergence" was adopted by authors with widely diverging philosophical views. It was particularly popular among vitalists, but for them, as is evident from the writings of Bergson and others, it was a metaphysical principle. This interpretation was shared by most of their opponents. J. B. S. Haldane (1932:113) remarked that "the doctrine of emergence ... is radically opposed to the spirit of science." The reason for this opposition to emergence is that emergence is characterized by three properties that appear at first sight to be in conflict with a straightforward mechanistic explanation: first, that a genuine novelty is produced – that is, some feature or process that was previously nonexistent; second, that the characteristics of this novelty are qualitatively, not just quantitatively, unlike anything that existed before; third, that it was unpredictable before its emergence, not only in practice, but in principle, even on the basis of an ideal, complete knowledge of the state of the cosmos.

The defenders of emergence insisted that this process should be considered simply an immanent property of nature, as documented by its universal occurrence. They point out that new properties may emerge whenever any more complex system is constructed from simpler components.

This was shown already by Mill and Lewes and was made more widely known by T. H. Huxley when referring to the emergence of the "aquosity" of water, a compound of two gases, hydrogen and oxygen. In the 1950s, Niels Bohr, who accepted emergence, also used water as an illustration of the principle of emergence. The emergence of unexpected properties at the molecular level, as in the case of the formation of water, demonstrates particularly persuasively that emergence is an empirical and not a metaphysical principle. This can also be demonstrated by another simple example, the emergence of the properties "hammer" when one brings handle and head together.

One of the standard objections of the reductionists to emergentism is that nothing new is produced in a case of emergence. But this claim is only a half-truth. To be sure, no new substance is produced; a hammer consists of the same substance as its separated components, handle and head. But something new has nevertheless been produced, the interaction of handle and head. Neither its wooden handle by itself nor the hammer head can perform (with any efficiency) the functions of a hammer. When one puts the two together, the properties of a hammer "emerge." And this newly added interaction *is* the crucial property of every emerged system, from the molecular level up. Emergence originates through the new relations (interactions) of the previously unconnected components. Indeed, not to take the importance of such connections into consideration is one of the basic failures of reductionism. The connection between hammer head and handle does not exist until the two are put together. The same is true for any interactions in a complex biological system. Dealing with the separated components tells us nothing about their interactions. And because these interactions in the living world are unique for every existing individual (except in asexual clones), their uniqueness refutes the claims of the reductionists.

For working scientists emergence of something qualitatively new is a daily encountered fact of life. They have no difficulty with this phenomenon because they know that the properties of higher systems are due not exclusively to the properties of the components but also to the ordering of these systems. Some authors have claimed that emergence is in conflict with Darwin's theory of gradual evolution because the new phenotype is separated by a distinct step. This objection, however, is due to a confusion of phenotypic gradualness and populational gradualness. What counts is that the evolutionary change takes place in populations, and a certain amount of discontinuity in the involved phenotypes is an irrelevant consideration.

It is now abundantly clear that evolutionary emergence is an empirical phenomenon without any metaphysical foundations. Acceptance of this principle is important because it helps to explain phenomena that previously had seemed to be in conflict with a mechanistic explanation of the evolutionary process. It eliminates any need to invoke metaphysical principles for the origin of novelties in the evolutionary process.

Theory reduction

Explanatory reductionism was not the only kind of reduction promoted by philosophers. Many of them supported a form of reduction called *theory reduction*. This form of reduction is based on the claim that theories and laws in one field of science are nothing but special cases of theories and laws formulated in some other, more basic branch of science, in particular of physical science. According to this belief, all regularities ("laws") observed in the living world are nothing but special cases of the laws and theories of the physical sciences. Therefore, to achieve the unification of science, it is the task of the philosopher of science to "reduce" the theories of biology to the more basic ones of the physical sciences.

Scientists, on the whole, showed little interest in theory reduction. It has been a concern mainly of the philosophers of science; indeed it is that aspect of reduction that was of the greatest interest to them (Hull

1972). The classical treatment is that of Nagel (1961). Theory reduction has been actively promoted also by Schaffner (1967, 1969) and Ruse (1971, 1973, 1976) and rather more cautiously by Rosenberg (1985). Decisive refutations have been provided by Hull (1974), Kitcher (1984), and Kincaid (1990).

The procedure of theory reduction is usually given as follows: "a theory T_2 (concerning a high level of organization) is reduced to a theory T_1 (concerning a lower level), if T_2 contains no primitive terms of its own, i.e., if the conceptual apparatus of T_1 is sufficient to express T_2." To state the conditions of strong reduction more concisely (Ayala 1968), to reduce a more special branch of science to a more basic one, it must be shown [according to Nagel (1961)]

(1) That all the laws and theories of the more specialized science are the logical consequences of the theoretical constructs of the more basic one; this is the condition of *derivability*.

(2) To accomplish this reduction, all technical terms used in the more specialized science must be redefinable in the terms of the more basic science; this is the condition of *connectability*.

The postulate of connectability encounters particular difficulties for the reduction of biological theories because the conceptual framework of biology is so totally different from that of the physical sciences that there is hardly ever any possibility for translating a biological term into one of physics or chemistry. Going through the glossaries of books in various branches of biology one encounters hundreds if not thousands of such untranslatable biological terms. Examples are territory, speciation, female choice, founder principle, imprinting, parental investment, meiosis, competition, courtship, and struggle for existence, to give only a few examples. This untranslatability of biological concepts was already known to Woodger (1929:263). Later it was particularly Beckner (1959) who called attention to it and listed many examples.

Reductionist philosophers usually have tried to support their case in favor of reduction by attempting to reduce Mendelian genetics to molecular genetics. But both Hull (1974) and particularly Kitcher (1984) have shown how unsuccessful this endeavor has been. It is not only the untranslatability of biological terms and concepts that makes theory reduction impossible but also the fact that very few biological generalizations can be connected with any of the laws of physics or chemistry. One particular difficulty is posed by the scarcity of laws relating to complex biological systems. Seeing all this evidence, Popper (1974:269, 279, 281) concluded "as a philosophy, reductionism is a failure . . . we live in a universe of emergent novelty; of a novelty which, as a rule, is not completely reducible to any of the preceding stages."

It is only in the biology of proximate causations that theory reduction is occasionally feasible. On the other hand, no principle of historical evolutionary theory can ever be reduced to the laws of physics or chemistry. Contrary to the claims of some reductionists, this has nothing to do with any alleged immaturity of biology. Indeed, the new insights gained by molecular genetics during the last forty years have made the impossibility of reduction even clearer than it was before (Kitcher 1984).

Consequences of the failure of reductionism

It is not so many years ago that Ruse asked "why many of today's great biologists are adamantly opposed to any kind of biological reductionist thesis?" (1973:217). The answer is now obvious. It is because these biologists understood the nature of the biological problems so much better than the physicalists who at that time dominated the philosophy of science. The popularity of reductionism sharply declined in the philosophy of science after its nature was better understood and, particularly, how it differed from analysis.

Reduction and philosophy

My treatment of reduction is that by a scientist. Philosophers of science would deal with the subject very differently, basing their arguments on laws, logic, and the equipment of the philosophy of science. A typical example of such an approach is Rosenberg's (2001) "Reductionism in a Historical Science." Most scientists fail to see what such a "philosophical" treatment would add to an understanding of a phenomenon or process. Reduction, by failing to consider the interaction of components, fails to fulfill what it promises. It can be ignored in the construction of any philosophy of biology.

LITERATURE CITED

Ayala, F. 1968. Biology as an autonomous science. *American Scientist*, 56:207–221.

Ayala, F., and T. Dobzhansky (eds.). 1974. *Studies in the Philosophy of Biology: Reduction and Related Problems*. Berkeley: University of California Press.

Beckner, M. 1959. *The Biological Way of Thought*. New York: Columbia University Press.

Dobzhansky, T. 1968. On Cartesian and Darwinian aspects of biology. *Graduate Journal*, 8(1):99–117.

Goudge, T. A. 1965. Another look at emergent evolutionism. *Dialogue*, 4(3):273–285.

Haldane, J. B. S. 1932. *The Causes of Evolution*. New York: Longmann, Green.

Hoyningen-Huene, P. 1989. Epistemological reductionism in biology. In *Reductionism and Systems Theory in the Life Sciences*, P. Hoyningen-Huene and F. M. Wuketis (eds.). Dordrecht: Kluwer, pp. 29–44.

Hull, D. 1969a. The natural system and the species problem. In *Systematic Biology*, C. G. Sibley (ed.). Washington, DC: National Academy Press, pp. 56–61.

Hull, D. 1969b. What philosophy of biology is not. *Journal of Historical Biology*, 2:241–268.

Hull, D. 1972. Reduction in genetics – biology or philosophy. *Philosophy of Science*, 39:491–499.

Hull, D. 1974. *The Philosophy of Biological Science*. Englewood, NJ: Prentice-Hall.

Kincaid, H. 1990. Molecular biology and the unity of science. *Philosophy of Science*, 57:575–593.

Kitcher, P. 1984. 1953 and all that. *Philosophical Reviews*, 93:335–373.

Lewes, G. H. 1874–1875. *Problems of Life and Mind*. 2 vols. London: Longmann, Green.

Lewontin, R. 1969. The bases of conflict in biological explanation. *J. Hist. Biol.*, 2:35–45.

Lorenz, K. 1973. *Die Rückseite des Spiegels*. München: R. Piper.

Mandelbaum, M. 1971. *History, Man and Reason*. Baltimore: Johns Hopkins Press.

Mayr, E. 1982. *The Growth of Biological Thought*. Cambridge, MA: Harvard University Press.

Mayr, E. 1988. The limits of reductionism. *Nature*, 331:475.

Morgan, C. L. 1923. *Emergent Evolution*. London: Williams and Norgate.

Nagel, E. 1961. *The Structure of Science. Problems in the Logic of Scientific Explanation*. New York: Harcourt, Brace, and World.

Popper, K. 1974. *Unended Quest*. La Salle, IL: Open Court Publishing.

Putnam, H. 1973. Reductionism and the nature of psychology. *Cognition*, 2:135.

Rosenberg, A. 1985. *The Structure of Biology Science*. Cambridge: Cambridge University Press.

Rosenberg, A. 2001. Reductionism in a historical science. *Philosophy of Science*, 68:135–168.

Ruse, M. 1971. Reduction, replacement, and molecular biology. *Dialectica*, 25:39–72.

Ruse, M. 1973. *The Philosophy of Biology*. London: Hutchinson.

Ruse, M. 1976. Reduction in genetics. *Boston Studies in Philosophy of Science*, Vol. 32, R. S. Cohen, et al. (eds.). Dordrecht: Reidel, pp. 633–651.

Schaffner, K. S. 1967. Approaches to reductionism. *Philosophy of Science*, 34:137–147.

Schaffner, K. S. 1969. Theories and explanations in biology. *Journal of the History of Biology*, 2:19–33.

Simpson, G. G. 1964. *This View of Life*. New York: Harcourt, Brace, and World.

Simpson, G. G. 1974. The concept of progress in organic evolution. *Social Research*, pp. 28–51.

Sterelny, K., and P. J. Griffith. 1999. *Sex and Death*. Chicago: University of Chicago Press.

Wald, G. 1963. *Molecular Biology at Harvard*. Newsletter. Harvard Foundation Advanced Study Research. 15 March 1963:1.

Woodger, J. H. 1929. *Biological Principles*. London: Routledge, Kegan, Paul.

5

Darwin's Influence on Modern Thought[1]

EVERY PERIOD IN THE HISTORY of civilized humans was dominated by a definite set of ideas or ideologies. This is as true for the ancient Greeks as for Christianity, the Renaissance, the Scientific Revolution, the Enlightenment, and modern times. It is a challenging question to ask what the source is of the dominating ideas of the present era. One can ask this question also in different terms. For instance, which books have had the greatest impact on current thinking? Inevitably, the Bible would have to be mentioned in first place. Before 1989, when the bankruptcy of Marxism was declared, Karl Marx's *Das Kapital* would clearly have been in second place, and it is still a dominating influence in many parts of the world. Sigmund Freud has been in and out of favor. Albert Einstein's

---

[1] Revised version of Mayr (2001).

83

biographer Abraham Pais made the exuberant claim that Einstein's theories "have profoundly changed the way modern men and women think about the phenomena of inanimate nature." No sooner had Pais said this, though, than he recognized the exaggeration. "It would actually be better to say 'modern scientists' than 'modern men and woman,'" he wrote, because one needs schooling in the physicalist style of thought and mathematical techniques to appreciate Einstein's contributions. Indeed I doubt that any of the great discoveries in the physics of the 1920s had any influence whatsoever on the thinking of the average person. However, the situation is different with Darwin's *On the Origin of Species* (1859). No other book, except for the Bible, has had a greater impact on our modern thinking. I hope to be able to show that this evaluation is justified not only because Darwin more than anyone else was responsible for the acceptance of a secular explanation of the world but also because he revolutionized our thinking about the nature of this world in surprisingly many other ways.

The first Darwinian revolution

Thinking about the world, before Darwin, was dominated by physics. Even though living nature, from Buffon on, was increasingly important in the thinking of philosophers, it could not be properly organized until biology had become a recognized branch of science. And this happened not until the middle of the nineteenth century. It necessitated the acceptance of entirely new ideas, ideas coming from biology, and neither established science nor philosophy was quite ready to accept them. Their acceptance required an ideological revolution. And this, as it eventually turned out, was indeed a very drastic revolution. This revolution required more – and more drastic – modifications of the average person's world view than had occurred in previous centuries. The reason why this is usually overlooked is that Darwin traditionally is considered simply an evolutionist. He was that undoubtedly; and it was clearly Darwin who established secular science. In the 1860s the description "Darwinism" described someone who rejected a supernatural origin of the world and its changes. It did not require an acceptance of natural selection (Mayr 1991). The introduction of secular science was the first Darwinian revolution.

Darwin's contributions to a new Zeitgeist

By replacing divine with secular science, Darwin profoundly revolutionized the thinking of the nineteenth century. But Darwin's impact was not limited to evolution and the consequences of evolutionary thinking, including branching evolution (common descent) and humans' position in the universe (descent from the primates); it also included a whole series of new ideologies. In part, they were refutations of time-honored concepts like teleology; in part they were the introduction of entirely new concepts, like biopopulation. In the aggregate, they had a real revolutionary impact on the thinking of modern humans.

Evolution is such an obvious phenomenon for any student of nature that its almost universal rejection up to the middle of the nineteenth century is somewhat of a riddle. As the geneticist Dobzhansky so rightly said, "Nothing in biology makes sense except in the light of evolution," which is surely correct for all of nonfunctional biology. To be sure there have been proponents of evolution before Darwin, beginning with Buffon, and even a well-thought-out theory of evolution by Jean Baptiste Lamarck, but as late as 1859 all laypersons, and even almost all naturalists and philosophers, still accepted a stable, constant world. With evolution staring everybody in the face, why was it nevertheless on the whole so unacceptable up to 1859? What was it that prevented the acceptance of the seemingly obvious?

It is my considered conclusion that certain fundamental ideologies and concepts, the components of the early nineteenth-century Zeitgeist, were what prevented an earlier acceptance of evolutionism. Let me now discuss some of these factors.

Secular science

A literal acceptance of every word in the Bible was the standard view of every orthodox Christian in the early nineteenth century. Everything in this world, as we see it, was created by God. Natural theology added the conviction that at the time of creation God had also instituted a set of laws that would continue to maintain the perfect adaptation of a well-designed world. Darwin challenged all three major components of this belief. He claimed, first, that the world is evolving rather than remaining constant; second, that new species are not specially created but derived from common ancestors; and third, that the adaptation of each species is continuously regulated by the process of natural selection. In Darwin's theories, there is no need for divine interference or the action of supernatural forces in the whole process of the evolution of the living world, and particularly none in the whole process of natural selection. Darwin's revolutionary proposal was, thus, to replace the divinely controlled world by a strictly secular world, run according to the natural laws.

Amazingly, Darwin's proposal of an evolving world owing to common descent was accepted after 1859 almost at once by the greater majority of naturalists and philosophers. This was true not only for England but also broadly for the continent, particularly for the German-speaking countries and for Russia. Almost overnight, the idea of evolution had become acceptable even though the controversy over the causes of evolution continued for another eighty years. Darwin himself was largely responsible for the rapidity of this shift, owing to the overwhelming amount of evidence for evolution presented in the *Origin*. Indeed, Darwin had done even more, and this usually is not mentioned in the Darwin biographies. He presented some fifty or sixty biological phenomena easily explained by natural selection but quite impervious to any explanation under special creation, and equally inexplicable to so-called intelligent design [see Darwin (1859: pp. 35, 95, 133, 139, 186, 188, 194, 203, 399, 406, 413, 420, 435, 456, 469, 478, 486, and many other nearby pages)].

Common descent and humans' position

Darwin's theory of common descent was so rapidly accepted because it supplied an explanation for the Linnaean hierarchy of kinds of organisms and for the findings of the comparative anatomists. However, the theory of common descent also led to one conclusion that was quite unpalatable to most of Darwin's Victorian contemporaries. It postulated that human ancestors were apes. If the humans had descended from apes, then they were not outside the rest of the living world but were actually part of it. This was the end of any strictly anthropomorphic philosophy. Even though Darwin did not question the unique characteristics of *Homo sapiens*, and neither do the modern evolutionists, nevertheless, zoologically humans are nothing but a specially evolved ape. Indeed, all modern investigations have revealed the incredible similarity between humans and chimpanzees. We share 98% of our genes, and many of our proteins — for instance, hemoglobin — are identical. It has become obvious in recent years that, in a philosophical study of humans, dealing with such questions as the nature of consciousness, intelligence, and human altruism, one can no longer ignore the origin of these human capacities in our anthropoid ancestors. This is true even though, through evolution, mankind has acquired many unique characteristics and capacities.

Population thinking

Let us now turn directly to an analysis of the philosophical foundations of Darwin's theorizing. With evolution so obvious to any student of living nature, why did it take so long before this obvious fact became acceptable? Let us study this for a particular case. Darwin's most original and most important new concept was that of natural selection. Why were not only the philosophers, but even most biologists, so hostile to this theory for such a long time? It is my claim that the conceptual framework of the period and, in particular, the almost universal acceptance of typological thinking – what Popper called essentialism – was responsible for this delay. This kind of thinking was first introduced into philosophy by Plato and the Pythagoreans, who postulated that the world consisted of a limited number of classes of entities (*eide*) and that only the type (essence) of each of these classes of objects had reality, all the seeming variations of these types being immaterial and irrelevant. The Platonian types (or *eide*) were considered to be constant, timeless, and sharply delimited against other such types. Such typological thinking was universally adopted by the physical scientists because all the fundamental entities of matter, such as the nuclear particles and the chemical elements, are indeed constant and sharply delimited against each other.

Darwin rejected such a description for organic diversity. Instead he introduced a mode of thinking we now refer to as *population thinking*. No two individuals in a biopopulation, not even identical twins, are actually identical. This is true even for the six billion individuals of the human species. It is this variation among the uniquely different individuals that has *reality*, while the calculated statistical mean value of this variation is an abstraction. This view was a totally new philosophical concept, crucial for the understanding of the theory of natural selection. How novel this concept was, appeared when Darwin himself sometimes slipped back into typological thinking. This was the reason he failed to solve the problem of the origin of new species.

Population thinking is of tremendous importance in daily life. For instance, the failure to apply population thinking is the major source of racism. Many of Darwin's associates, such as Charles Lyell and T. H. Huxley (Mayr 1982), never adopted population thinking and remained typologists all their lives. Consequently they were unable to understand and accept natural selection. Typological thinking was so firmly rooted in the thinking of the period that it is not surprising it took eighty years until, in the 1930s, the concept of natural selection was finally universally adopted by evolutionists.

The genetic program

It was Darwin who contributed the concept of biopopulation, one of the fundamental differences between the living and the inanimate world. Another one, equally exclusive to the living world, the *genetic program*, could not be conceived until cytology, genetics, and molecular biology had matured. It is responsible for the dual causation of all activities of and in living organisms.

Perhaps the most profound difference between the inanimate world of the physicist and the living world of the biologist is the dual causation of all organisms. Anything and everything that happens in the physical world is exclusively controlled by the natural laws, gravitation, the thermal laws, and the scores of other natural laws discovered by the physical sciences. These laws describe the properties of all matter, and even living organisms and their parts are, as matter, as subject as inanimate matter to these laws. The laws of the physical sciences are particularly evident in the study of life at the cellular and molecular level. Theory formation in physiology is based almost exclusively on natural laws. However, organisms are subject also to a second set of causal factors, the information provided by their genetic program. There is no activity, movement, or behavior of an organism that is not influenced by the genetic program. This program, consisting of the genotype of each living individual, is the product of billions of years of natural selection in every generation. The structural laws and the messages from the genetic program function simultaneously and in harmony, but genetic programs occur only in living organisms. They provide an absolute borderline between the inanimate and the living world.

Naturalists, of course, have been aware of this fundamental difference for thousands of years, but their explanation for it was invalid. They tried to attribute life to the occult force of vitalism, a *vis vitalis*, but eventually it was determined that such a force does not exist. Darwin was not a vitalist, but he was not able to explain life. This was finally made possible in the twentieth century by the discoveries of cytology, genetics, and molecular biology. The sciences finally provided us with a naturalistic explanation of life.

Finalism

Let me now turn to another dominant concept in philosophy in the first half of the nineteenth century. When philosopher Immanuel Kant, in his *Critique of Judgment* (1790), tried to develop a philosophy of biology on the basis of the physicalist philosophy of Newton, he failed embarrassingly. Finally he concluded that biology is different from the physical sciences and that we must find some philosophical factor not used by Newton. Indeed, he thought he had found such a factor in Aristotle's fourth cause, the final cause (teleology). And so Kant ascribed to teleology not only evolutionary change (not really recognized by him as such) but also everything else in biology that he was not able to explain by Newtonian laws. This had a rather adverse effect on German nineteenth-century philosophy, because an unsupported reliance on teleology played an important role in the philosophies of all of Kant's followers.

It was Darwin's great achievement to be able to explain by natural selection all the phenomena for which Kant had thought he needed to invoke teleology. The great American philosopher Willard Van Ormond Quine, in a conversation I had with him about a year before his death, told me that he considered Darwin's greatest philosophical achievement to consist in having refuted Aristotle's final cause. The purely automatic process of natural selection, producing abundant variation in every generation and always removing the inferior individuals and favoring the best adapted ones, can explain all processes and phenomena that, before 1859, could be explained only by teleology. At the present we still recognize four teleological phenomena or processes in nature (see chapter 3) but they can all be explained by the laws of chemistry and physics, while a cosmic teleology, such as that adopted by Kant, does not exist.

The role of chance

Determinism was a ruling philosophy before Darwin. As Laplace had boasted, if he knew the exact location and motion of every object in the universe, then he would be able to predict every detail of the future history of the world. There was no room in his philosophy for chance or accident. Darwin also paid strict lip service to such determinism. He accepted the standard belief of his period that every chance process in the universe had a cause. But the Newtonian laws of physics were not sufficient to explain genetic variation. So Darwin made use of the then universally accepted principle of an inheritance of acquired characters. Domestic animals, he said, are more variable than wild ones because they have a richer diet and the changes thus produced are inherited. For him all mutations were the result of an observable cause. It was not until the

1890s that the concept of spontaneous mutations was introduced into biology by DeVries.

Darwinian variation, not being based on Newtonian natural laws, was not acceptable to the contemporary philosophers. Such variants were considered chance phenomena or accidents. The physicist-philosopher Herschel referred to natural selection contemptuously as the law of the higgledy-piggledy. He was not alone in this criticism; the Cambridge geologist Sedgwick and other critics of Darwin chided him for invoking chance as an evolutionary factor. Again and again Darwin was asked, how can you believe that such a perfect organ as the eye originated by chance? We still lack a thorough analysis of the history of the gradual acceptance of chance in scientific explanation. Now that it is realized that chance in evolution is part of the two-step nature of the process of natural selection (chapter 7), the processes of selection or elimination during the second step of natural selection can make use of the positive contribution made by random variation at the first step.

At about the same time, the middle of the nineteenth century, the importance of chance was also discovered in the physical sciences, and Darwin's sponsorship of chance soon was no longer criticized so severely. When modern authors speak of chance variation, they do not deny the existence of molecular causal forces, but they deny the claim that such genetic variation is a response to the adaptive needs of an organism. Such a response never occurs, and molecular biology has shown that there is no inheritance of acquired characters. In spite of his uncertainties, Darwin certainly was one of the great pioneers in making the chance nature of many biological phenomena an acceptable concept.

Laws

Theories in the Newtonian philosophy of science usually were based on laws. Darwin on the whole accepted this view. And so we find that he uses the term "law" very freely in the *Origin*. Any cause or event that seemed to occur at all regularly was called by him a law. However, I rather agree with those modern philosophers who deny the legitimacy of referring to evolutionary regularities as laws, because these regularities do not deal with the basics of matter as do the laws in physics. They are invariably restricted in space and time, and they usually have numerous exceptions. This is why Popper's falsification principle usually cannot be applied in evolutionary biology, because exceptions do not falsify the general validity of most regularities.

If one concludes that there are no natural laws in evolutionary biology, one must ask, on what can one then base biological theories? The view now widely adopted is that theories in evolutionary biology are based on *concepts* rather than on laws, and this branch of science certainly has abundant concepts on which to base theories. Let me just mention concepts such as natural selection, struggle for existence, competition, biopopulation, adaptation, reproductive success, female choice, and male dominance. I admit that some of these concepts, with a little effort, perhaps can be converted into pseudo-laws, but there is no question that such "laws" are something very different from the Newtonian natural laws. As a result, a philosophy of physics based on natural laws turns out to be something very different from a philosophy of biology based on concepts.

Darwin himself was quite unaware of this difference, although it was he, perhaps more than anyone else, who introduced the new practice of theory formation on the basis of concepts rather than of natural laws.

Darwin's method

Darwin was first and foremost a naturalist. His favorite method was also that of the naturalist; he made a series of observations and developed conjectures from this evidence. He considered this approach to be the inductive method and recorded in his autobiography that he considered himself a true follower of Bacon. However, some students of Darwin's work – for instance, Ghiselin (1969) – thought this approach was better considered to be hypothetico-deductive. Actually, perhaps the closest to the truth would be to say that Darwin was a pragmatist and used whatever method he thought would bring him the best results. Darwin was a very keen observer, and there is no doubt that observation was his most productive approach. However, he was also a skillful experimenter and, particularly in his botanical researches, he conducted numerous experiments. Like all naturalists, the method he used perhaps most frequently was the comparative method.

Time

The most widely used method in the physical sciences is the experiment. However, in his evolutionary studies Darwin had to cope with a factor that is irrelevant in most of the physical sciences except in geology and cosmology, the time factor. One cannot experiment with biological happenings in the past. Phenomena like the extinction of the dinosaurs and all other evolutionary events are inaccessible to the experimental method and require an entirely different methodology, that of the so-called historical narratives. In this method one develops an imaginary scenario of past happenings on the basis of their consequences. One then makes all sorts of predictions from this scenario and determines whether they have come true. Darwin used this method particularly successfully in his biogeographic reconstructions. Which postulated former land bridges, for instance, are supported by current distributions and which are not?

The importance of the method of historical narratives has long been overlooked by philosophers. It is, however, an indispensable method whenever one deals with the consequences of past events. Considering the productivity of this method, it is surprising how much it has been neglected by the historians of science. How much use, for instance, have Buffon, Linnaeus, Lamarck, and Blumenbach made of historical narratives?

In my writings I have referred to the philosophical foundations of Darwin's thought, and I have called Darwin one of the great philosophers. This is not a widely adopted point of view. Even though he was one of the great philosophers of all time, his philosophy of biology differs so fundamentally from the philosophies based on logic, mathematics, and the physical sciences that its philosophical nature was traditionally overlooked.

Summary

Let me now try to summarize Darwin's contributions to the thinking of modern humans. He was responsible for the replacement of a world view based on Christian dogma by a strictly secular world view. Furthermore, his writings led to the rejection of several previously dominant world views such as essentialism, finalism, determinism, and the sufficiency of Newtonian laws for the explanation of evolution. He replaced these refuted concepts with a number of new ones of wide-reaching importance also outside of biology, such as biopopulation, natural selection, the importance of chance and contingency, the explanatory importance of the time factor (historical narratives), and the importance of the social group for the origin of ethics. Almost every component in modern human's belief system is somehow affected by one or another of Darwin's conceptual innovations. His opus as a whole is the foundation of a rapidly developing new philosophy of biology. There can be no doubt that the thinking of every modern Western person has been profoundly affected by Darwin's philosophical thought.

LITERATURE CITED

Darwin, C. 1859. *On the Origin of Species by Means of Natural Selection or the Preservation of Favored Races in the Struggle for Life*. London: John Murray [1964, Facsimile of the First Edition; Cambridge, MA: Harvard University Press].

Ghiselin, M. 1969. *The Triumph of the Darwinian Method*. Berkeley: University of California Press.

Kant, I. 1755. *Allgemeine Naturgeschichte und Theorie des Himmels* [*General Natural History and Theory of the Sky*].

Kant, I. 1790. *Die Kritik der Urteilskraft* [*Critique of Judgment*]. Berlin: Georg Reimer.

Lloyd, E. A. 1988. *The Structure and Confirmation of Evolutionary Theory*. Contributions in Philosophy, Vol. 37. New York: Greenwood.

Lovejoy, A. O. 1936. *The Great Chain of Being*. Cambridge, MA: Harvard University Press.

Mayr, E. 1959. Darwin and the evolutionary theory in biology. In *Evolution and Anthropology. A Centennial Appraisal*, B. J. Meggers (ed.). Washington, DC: The Anthropological Society of Washington, pp. 1–10.

Mayr, E. 1982. *The Growth of Biological Thought*. Cambridge, MA: Harvard University Press (Belknap Press).

Mayr, E. 1991. The ideological resistance to Darwin's theory of natural selection. *Proceedings of the American Philosophical Society*, 135:123–139.

Mayr, E. 1993. The resistance to Darwinism and the misconceptions on which it was based. In *Creative Evolution*, W. Schopf and J. Campbell (eds.). Boston: Jones and Bartlett, pp. 35–46.

Mayr, E. 2001. The philosophical foundations of Darwinism. *Proceedings of the American Philosophical Society*, 145:488–495.

6

Darwin's Five Theories of Evolution[1]

DARWIN WAS AN INVETERATE THEORIZER and became the author of numerous evolutionary theories, some big, some small. He usually referred to his evolutionary theories in the singular as "my theory" and treated the nonconstancy of species, common descent, and natural selection as a single, individual theory. I still remember how shocked I was as a young evolutionist when I discovered that Darwin had rejected Moritz Wagner's perfectly valid theory of geographic speciation (and the importance of isolation) because Wagner had not adopted natural selection. How could Charles Darwin, my hero, do something I considered totally illogical? His failure to recognize the independence of the various

[1] Greatly revised version of Mayr (1985).

97

Table 6.1. *Acceptance of Darwinian theories by evolutionists*

	Common descent	Gradualness	Populational speciation	Natural selection
Lamarck	No	Yes	No	No
Darwin	Yes	Yes	Yes	Yes
Haeckel	Yes	Yes	?	In part
Neo-Lamarckians	Yes	Yes	Yes	No
T. H. Huxley	Yes	No	No	(No)
De Vries	Yes	No	No	No
T. H. Morgan	Yes	(No)	No	Unimportant

theories of his evolutionary paradigm also caused Darwin difficulties in his discussion of the principle of divergence (Mayr 1992). I have recently come to the conclusion that Darwin's blindness to recognize this became one of the main reasons for the never-ending controversies on evolutionary biology after 1859. However, by now, it has become quite clear that Darwin's paradigm consists of several major independent theories (Mayr 1985). Not surprisingly, different evolutionists disagreed with each other about the validity of these theories and founded opposing schools. These were feuding with each other for almost eighty years until a synthesis was achieved in the 1930s and 1940s.

An analysis of the manifold Darwinian theories led me to the conclusion that Darwin's paradigm consists of five major independent theories. That these theories are indeed "logically independent" of each other has been confirmed by several recent authors. The acceptance of some and at the same time rejection of different theories of this package of five theories by particular authors is perhaps the best evidence for their independence (Table 6.1).

I present here only an abridged (but somewhat revised) version of a very detailed analysis of Darwin's five theories (Mayr 1985). I refer to this monographic treatment for additional information on these five theories.

There is one particularly cogent reason why Darwinism cannot be a single homogeneous theory: Organic evolution consists of two essentially independent processes, transformation in time and diversification in (ecological and geographic) space. The two processes require a minimum of two entirely independent and very different theories. When writers on Darwin have nevertheless almost invariably spoken of the combination of these various theories as "Darwin's theory" in the singular, it was largely Darwin's own doing. He not only referred to the theory of evolution itself as "my theory," but he also called the theory of common descent by natural selection "my theory," as if common descent and natural selection were a single theory.

The discrimination among his various theories was not helped by the fact that Darwin treated speciation under natural selection in chapter 4 of the *Origin* and that he ascribed many phenomena, particularly those of geographic distribution, to natural selection when they were really the consequences of common descent. Under the circumstances, I consider it urgently necessary to dissect Darwin's conceptual framework of evolution into the major theories that formed the basis of his evolutionary thinking. For the sake of convenience I have partitioned Darwin's evolutionary paradigm into five theories; of course, others might prefer a different division. When later authors referred to Darwin's theory they invariably had a combination of some of the following five theories in mind. For Darwin himself these five theories were evolution as such, common descent, gradualism, multiplication of species, and natural selection. Someone might claim that indeed these five theories are a logically inseparable package and that Darwin was quite correct in treating them as such. This claim, however, is refuted by the fact, as I have demonstrated elsewhere (Mayr 1982b:505–510), that most evolutionists in the immediate post-1859 period – that is, authors who had accepted the theory of nonconstancy of species – rejected one or several of Darwin's other four theories. This demonstrates that the five theories are not one indivisible whole.

Evolution as such

This is the theory that the world is neither constant nor perpetually cycling but instead is steadily and in part directionally changing and that organisms are being transformed in time. It is difficult for a modern to visualize how widespread the belief still was in the first half of the nineteenth century, particularly in England, that the world is essentially constant and of short duration. Even most of those who, like Charles Lyell, were fully aware of the great age of the earth and of the steady march of extinction, refused to believe in a transformation of species. A belief in evolution was also referred to as the theory of the nonconstancy of species.

Evolution as such is no longer a theory for a modern author. It is as much a fact as that the earth revolves around the sun rather than the reverse. The changes documented by the fossil record in precisely dated geological strata are the fact that we designate as evolution. It is the factual basis on which the other four evolutionary theories rest. For instance, all the phenomena explained by common descent would make no sense if evolution were not a fact.

Common descent

The case of the three species of Galapagos mockingbirds provided Darwin with an important new insight. The three species had clearly descended from a single ancestral species on the South American continent. From this conclusion it was only a small step to postulate that all mockingbirds were derived from a common ancestor – indeed, that every group of organisms descended from an ancestral species. This is Darwin's theory of common descent.

It must be emphasized that the two terms *common descent* and *branching* describe exactly the same phenomenon for an evolutionist. Common descent reflects a backward-looking view and branching a forward-looking view. The concept of common descent was not entirely original with Darwin. Buffon had already considered it for close relatives, such as horses and asses; but, not accepting evolution, he had not extended this thought systematically. There are occasional suggestions of common descent in a number of other pre-Darwin writers, but historians so far have not made a careful search for early adherents of common ancestry. It is a theory that definitely was not upheld by Lamarck, who, although he proposed the occasional splitting of "masses" (higher taxa), never thought in terms of a splitting of species and regular branching. He derived diversity from spontaneous generation and the vertical transformation of each line separately into stages of higher perfection. For him descent was linear descent within each phyletic line, and the concept of common descent was alien to him.

None of Darwin's theories was accepted as enthusiastically as common descent; it is probably correct to say that no other of Darwin's theories had such enormous immediate explanatory powers. Everything that had seemed to be arbitrary or chaotic in natural history up to that point now began to make sense. The archetypes of Owen and of the comparative anatomists could now be explained as the heritage from a common ancestor. The entire Linnaean hierarchy suddenly became quite logical, because it was now apparent that each higher taxon consisted of the descendants of a still more remote ancestor. Patterns of distribution that previously had seemed capricious could now be explained in terms of the dispersal of descendants. Virtually all the proofs for evolution listed by Darwin in the *Origin* actually consist of evidence for common descent. To establish the line of descent of isolated or aberrant types became the most popular research program of the post-*Origin* period and has largely remained the research program of comparative anatomists and paleontologists almost up to the present day. To shed light on common ancestors also became the program of comparative embryology. Even those who did not believe in strict recapitulation often discovered similarities in embryos that were obliterated in the adults. These similarities, such as the chorda in tunicates and vertebrates, and the gill arches in fishers and terrestrial tetrapods, had been totally mystifying until they were interpreted as vestiges of a common past.

Nothing helped the rapid adoption of evolution more than the explanatory power of the theory of common descent. Soon it was demonstrated that even animals and plants, seemingly so different from each other, could be derived from a common, one-celled ancestor. This Darwin had already predicted when he suggested that "all our plants and animals [have descended] from some one form, into which life was first breathed" (*Natural Selection*, p. 248). The studies of cytology (meiosis, chromosomal inheritance) and biochemistry fully confirmed the evidence from morphology and systematics for a common origin. It was one of the triumphs of molecular biology to be able to establish that eukaryotes and prokaryotes have identical genetic codes, thus leaving little doubt about the common origin even of these groups. Even though there are still a number of connections to be established among higher taxa, particularly among the phyla of plants and invertebrates, there is probably no biologist left today who would question that all organisms now found on the earth have descended from a single origin of life.

There was only one area in which the application of the theory of common descent encountered vigorous resistance: the inclusion of humans into the total line of descent. To judge from contemporary cartoons, none of the Darwinian theories was less acceptable to the Victorians than the derivation of humans from the other primates. Yet at the present time this derivation is not only remarkably well substantiated by the fossil record, but the biochemical and chromosomal similarity of humans and the African apes is so great that it is quite puzzling why they are so relatively different in morphology and brain development.

Gradualism versus saltationism

Darwin's third theory was that evolutionary transformation always proceeds gradually and never in jumps. One will never understand Darwin's insistence on the gradualism of evolution, nor the strong opposition to this theory, unless one realizes that virtually everyone at that time was an essentialist. The occurrence of new species, documented by the fossil record, could take place only by new origins – that is, by saltations. However, because the new species were perfectly adapted and because there was no evidence for a frequent production of maladapted species, Darwin saw only two alternatives. Either the perfect new species had been specially created by an all-powerful and all-wise Creator, or else – if such a supernatural process were unacceptable – the new species had evolved gradually from preexisting species by a slow process, at each stage of which they maintained their adaptation. It was this second alternative that Darwin adopted.

This theory of gradualism was a drastic departure from tradition. Theories of a saltational origin of new species had existed from the pre-Socratics to Maupertuis and the progressionists among the so-called catastrophist geologists. These saltationist theories were consistent with essentialism.

Darwin's totally gradualist theory of evolution – not only species but also higher taxa arise through gradual transformation – immediately encountered strong opposition. Even Darwin's closest friends were unhappy about it. T. H. Huxley wrote to Darwin on the day before publication of the *Origin*: "You have loaded yourself with an unnecessary difficulty in adopting *Natura non facit saltum* so unreservedly . . ." (Darwin F. 1887: 2, 27). In spite of the urgings of Huxley, Galton, Kölliker, and other contemporaries, Darwin insisted almost obstinately on the gradualness of evolution, even though he was fully aware of the revolutionary nature of this concept. With the exceptions of Lamarck and Geoffroy, almost everybody else who had ever thought about changes in the organic world had been an essentialist and had resorted to saltations.

The source of Darwin's strong belief in gradualism is not quite clear. The problem has not yet been analyzed adequately. Most likely gradualism is an extension of Lyell's uniformitarianism from geology to the organic world. Lyell's failure to do so had rightly been criticized by Bronn. Darwin, of course, also had strictly empirical reasons for his insistence on gradualism. His work with domestic races, particularly his work with pigeons and his conversations with animal breeders, convinced him how strikingly different the end products of slow, gradual selection could be. This fitted well with his observations on the Galapagos mockingbirds and tortoises, which were best explained as the result of gradual transformation.

Finally, Darwin had didactic reasons for insisting on the slow accumulation of rather small steps. He answered the argument of his opponents that one should be able to "observe" evolutionary change owing to natural selection by saying: "As natural selection acts solely by accumulating slight successive favorable variations, it can produce no great or sudden modifications; it can act only by very short and slow steps" (*Origin*, p. 471). There is little doubt that the general emergence of population thinking in Darwin strengthened his adherence to gradualism. As soon as one adopts the concept that evolution occurs in populations and slowly transforms them – and this is what Darwin increasingly believed – one is automatically forced also to adopt gradualism. Gradualism and population thinking probably were originally independent strands in Darwin's conceptual framework, but eventually they reinforced each other powerfully.

The naturalists were the main supporters of gradual evolution, which they encountered everywhere in the form of geographic variation. Eventually, geneticists arrived at the same conclusion through the discovery of ever slighter mutations, of polygeny, and of pleiotropy. The result was that gradualism was able to celebrate a complete victory during the evolutionary synthesis in spite of the continuing opposition by Gold-schmidt and Schindewolf.

Defining gradualism as populational evolution – and this is what Darwin basically had in mind – permits us to say that, in spite of all the opposition to him, Darwin ultimately prevailed even with his third evolutionary theory. Among the clearly established exceptions to gradualism are cases of stabilized hybrids that can reproduce without crossing (like allotetraploids) and cases of symbiogenesis (Margulis and Sagan 2002).

Nothing is said in the theory of gradualism about the rate at which the change may occur. Darwin was aware that evolution could sometimes progress quite rapidly, but, as Andrew Huxley (1981) has recently quite rightly pointed out, it could also contain periods of complete stasis "during which these same species remained without undergoing any change." In his well-known diagram in the *Origin* (opposite p. 117), Darwin lets one species (F) continue unchanged through 14,000 generations or even through a whole series of geological strata (p. 124). The understanding of the independence of gradualness and evolutionary rate is important for the evaluation of the theory of punctuated equilibria (Mayr 1982c).

The multiplication of species

This theory of Darwin's deals with the explanation of the origin of the enormous organic diversity. It is estimated that there are five to ten million species of animals and one to two million species of plants on earth. Even though only a fraction of this number were known in Darwin's day, the problem of why there are so many species and how they originated was already present. Lamarck had ignored the possibility of a multiplication of species in his *Philosophie Zoologique* (1809). For him, diversity was produced by differential adaptation. New evolutionary lines originated by spontaneous generation, he thought. In Lyell's steady-state world, species number was constant and new species were introduced to replace those that had become extinct. Any thought of the splitting of a species into several daughter species was absent among these earlier authors.

To find the solution to the problem of species diversification required an entirely new approach, and only the naturalists were in a position to find it. L. von Buch in the Canary Islands, Darwin in the Galapagos, Wagner in North Africa, and Wallace in Amazonia and the Malay Archipelago were the pioneers in this endeavor. By adding the horizontal (geography) to the vertical dimension that had previously monopolized evolutionary thought, they all were able to discover geographically representative (allopatric) species or incipient species. But more than that, these naturalists found numerous allopatric populations that were in all conceivable intermediate stages of species formation. The sharp discontinuity between species that had so impressed John Ray, Carl Linnaeus, and other students of the nondimensional situation (the local naturalists) was now supplemented by a continuity among species owing to the incorporation of the geographic dimension.

If one defines species simply as morphologically different types, one evades the real issue of the multiplication of species. A more realistic formulation of the problem of speciation was not possible until the development of the biological species concept (K. Jordan, Poulton, Stresemann, Mayr). Only then was it seen that the real problem is the acquisition of reproductive isolation among contemporary species. Transformation of a phyletic line in the time dimension (gradual phyletic evolution, as it was later designated) sheds no light on the origin of diversity. What then is it that does?

Darwin struggled with the problem of the multiplication of species all his life. Only after he had discovered the three new species of mockingbirds on different islands in the Galapagos did Darwin develop a fully consistent concept of geographic speciation. His thinking, at that period, seems to have been derived exclusively from the zoological literature.

But in due time Darwin became acquainted with varieties of plants, particularly through his friend the botanist Hooker (Kottler 1978,

Sulloway 1979). And this new information seemed to complicate the picture. What Darwin did not realize was that the botanists were using the term variety not for geographic races (subspecies), as did the zoologists, but for an entirely different kind of variants. For a botanist, more often than not, a variety was an individual variant ("morph") within a population. Because up to that time a variety [of animals] being a geographic race was an incipient species, Darwin assumed that the same was true for any variety, including those of plants. Hence, an individual variety of plants was an incipient species. Up to that expansion of Darwin's terminology from geographic variety to individual variety, speciation was a geographic process. However, if several individual varieties coexisting at the same locality could simultaneously become different new species, then speciation could be a sympatric process. And Darwin developed, assisted by his new "principle of divergence," a new scenario of sympatric speciation (Mayr 1992). And Darwin's scenario was apparently so convincing that, from the 1860s on, sympatric speciation, based on the principle of divergence, became as popular as geographic speciation, based on the isolation of geographic varieties (subspecies). The application of the principle of divergence to the process of speciation is a complex process and I refer for its explanation to a special analysis (Mayr 1992). Darwin's treatment of speciation in the *Origin* reveals his confusion about species and speciation. This was not cleared up until the synthesis in the 1940s.

Although Darwin deserves credit, together with Wallace, for having posed concretely for the first time the problem of the multiplication of species, the pluralism of his proposed solution led to a history of continuing controversy that is not entirely ended to this very day. At first, from the 1870s to the 1940s, sympatric speciation was perhaps the more popular theory of speciation, although some authors, particularly ornithologists and specialists of other groups displaying strong geographic variation, insisted on exclusive geographic speciation. Most entomologists, however, and likewise most botanists, even though admitting the occurrence of geographic speciation, considered sympatric speciation to be the more common and thus more important form of speciation. After 1942 allopatric speciation was more or less victorious for some twenty-five years, but then so many well-analyzed cases of sympatric speciation were found, particularly among fishes and insects, that there is now no longer any doubt about the frequency of sympatric speciation.

Paleontologists, on the whole, completely ignored the problem of the multiplication of species. For instance, one finds no discussion of it in the work of G. G. Simpson. The paleontologists finally incorporated speciation into their theories (Eldredge and Gould 1972), but their conclusions were based on the speciation research of those who studied living organisms.

There are three reasons why speciation is still such a problem 145 years after the publication of the *Origin*. The first is that, as in so much of evolutionary research, the evolutionists analyzes the results of past evolutionary processes and thus are obliged to reach their conclusions by inference. Consequently, one encounters all the well-known difficulties met in the reconstruction of historical sequences. The second difficulty is that, in spite of all the advances of genetics, we are still almost entirely ignorant about what happens genetically during speciation. And finally, it has become evident that rather different genetic mechanisms are involved in the speciation of different kinds of organisms and under different circumstances.

A rather unexpected discovery in the 1970s was responsible for the wide acceptance of sympatric speciation since the 1970s. As I pointed out in 1963, successful sympatric speciation is possible only if there is a simultaneous cooperation of two new factors, niche preference and mate preference. My former aversion to sympatric speciation was based on my assumption that these two preferences would be dealt with separately by natural selection. Recent research, particularly on the Cameroon cichlid fishes, however, showed that the two preferences could be combined. If, for instance, females would have a preference for males of a particular feeding niche – i.e., benthic feeder – and for males indicating this preference by their phenotype, this joint preference could quickly produce a new sympatric species. My assumption of a separate inheritance of the two preferences was invalid. No cases of sympatric speciation are known to me from mammals or birds. However, it is presumably frequent in host-specific groups of insects. Mapping the geographic ranges of closely related species in families such as the Cerambycidae and Buprestildae should provide an answer.

Natural selection

Darwin's theory of natural selection was his most daring, and most novel, theory. It dealt with the mechanism of evolutionary change and, more particularly, how this mechanism could account for the seeming harmony and adaptation of the organic world. It attempted to provide a natural explanation in place of the supernatural one of natural theology. Darwin's theory for this natural mechanism was unique. There was nothing like it in the whole philosophical literature from the pre-Socratics to Descartes, Leibniz, Hume, or Kant. It replaced teleology in nature with an essentially mechanical explanation.

I presented a detailed analysis of natural selection in chapter 5. To avoid duplication, I limit myself in this chapter to only a few aspects of selection. For Darwin, and for every Darwinian since, natural selection proceeds in two steps: the production of variation and the sorting of this variation by selection and elimination.

Although I call the theory of natural selection Darwin's fifth theory, it is actually, in turn, a small package of theories. This includes the theory of the perpetual existence of a reproductive surplus (superfecundity), the theory of the heritability of individual differences, the discreteness of the determinants of heredity, and several others. Many of these were not explicitly stated by Darwin but are implicit in his model as a whole. However, all of them are compatible with the populational nature of selection. All selection takes place in populations and changes the genetic composition of every population generation after generation. This is in complete contrast to the discontinuous character of saltational evolution by way of reproductively isolated individuals. What is invariably ignored, however, is that even continuous evolution is mildly discontinuous owing to the sequence of generations. In each generation an entirely new gene pool is reconstituted from which the individuals are drawn that are the targets of selection in that generation.

The theory of natural selection was the most bitterly resisted of all of Darwin's theories. If it were true, as some sociologists have claimed, that the theory was the inevitable consequence of the Zeitgeist of early nineteenth-century Britain, of the industrial revolution, of Adam Smith and the various ideologies of the period, one would think that the theory of natural selection would have been embraced at once by almost everybody. Exactly the opposite is true; the theory was almost universally rejected. In the 1860s only a few naturalists, like Wallace, Bates, Hooker, and Fritz Müller, could be called consistent selectionists. Lyell never had any use for natural selection, and even T. H. Huxley, defending it in public, was obviously uncomfortable with it and probably did not really believe in it (Poulton 1896, Kottler 1985). Before 1900 not a single experimental biologist either in Britain or elsewhere adopted the theory (Weismann was basically a naturalist). Of course, even Darwin was not a total selectionist, because he always allowed for the effects of use and disuse and for an occasional direct influence of the environment. The most determined resistance came from those who had been raised under the ideology of natural theology. They were quite unable to abandon the idea of a world designed by God and to accept instead a mechanical process. More importantly, a consistent application of the theory of natural selection meant a rejection of any and all cosmic teleology. Sedgwick and

K. E. von Baer were particularly articulate in resisting the elimination of teleology.

Natural selection represents not only the rejection of any finalistic causes that may have a supernatural origin, but it also rejects any and all determinism in the organic world. Natural selection is utterly "opportunistic," as G. G. Simpson has called it; it is a "tinkerer" (Jacob 1977). It starts, so to speak, from scratch in every generation, as I described above. Throughout the nineteenth century the physical scientists were still deterministic in their outlook, and so indeterministic a process as natural selection was simply not acceptable to them. One has only to read the critiques of the *Origin* written by some of the best-known physicists of the period (Hull 1973) to see how strongly the physicists objected to Darwin's "law of the higgledy-piggledy" (F. Darwin 1887:2, 37; Herschel 1861, p. 12). From the Greeks to the present day there has been a never-ending argument about whether the events of nature are due to chance or to necessity (Monod 1970). Curiously, in the controversies over natural selection, the process has often been described as "pure chance" (Herschel and many other opponents of natural selection) or as a strictly deterministic optimization process. Both classes of claimants overlook the two-step nature of natural selection and the fact that, in the first step, chance phenomena prevail, while the second step is decidedly of an antichance nature. As Sewall Wright so correctly said: "The Darwinian process of continued interplay of a random and a selective process is not intermediate between pure chance and pure determination, but in its consequences qualitatively utterly different from either" (1967, p. 117).

Even though everybody very soon accepted evolution, at first only a minority of biologists and very few nonbiologists became consistent selectionists. This was true until the period of the evolutionary synthesis. Instead, they adopted finalistic theories, neo-Lamarckian theories, and saltational theories. The controversy over natural selection is by no means at an end. Even today the relationship between selection and adaptation is hotly debated in the evolutionary literature, and it has been questioned whether it is legitimate to adopt an "adaptationist program" – that is, to search for the adaptive significance of the various characteristics of organisms (Gould and Lewontin 1979). But the question that is really before us is not so much whether natural selection is now universally adopted by evolutionists – a question one can unhesitatingly answer affirmatively – but rather whether the modern evolutionists' concept of natural selection is still that of Darwin or is considerably modified.

When Darwin first developed his theory of natural selection, he was still inclined to think that it was able to produce near-perfect adaptation, in the spirit of natural theology (Ospovat 1981). More thinking and the realization of the numerous deficiencies in the structure and function of organisms – perhaps particularly the incompatibility of a perfection-producing mechanism with extinction – led Darwin to reduce his claims for selection, so that all he demanded in the *Origin* was that "natural selection tends only to make each organism, each organic being, as perfect as, or slightly more perfect than, the other inhabitants of the same country with which it has to struggle for existence" (p. 201). Today we are even more conscious of the numerous constraints that make it impossible for natural selection to achieve perfection, or, to state it perhaps more realistically, to come even anywhere near to perfection (Gould and Lewontin 1979, Mayr 1982a).

The varying fates of Darwin's five theories

We can now summarize the subsequent fate of each of the five theories of Darwin, which I discussed above. Evolution as such, as well as the theory of common descent, was adopted very quickly. Within fifteen years of the publication of the *Origin*, hardly a qualified biologist was left who had not become an evolutionist. Gradualism, by contrast, had to struggle, because populational thinking was a concept that apparently was very difficult for anyone to adopt who was not a naturalist. Even today, in the discussions of punctuated equilibria, statements are made that indicate that some people still do not understand the core of population thinking. What counts is not the size of the individual mutation but only whether the introduction of evolutionary novelties proceeds through their gradual incorporation into populations or through the productions of a single new individual that is the progenitor of a new species or higher taxon.

That a theory of the multiplication of species is an essential, in fact integral, component of evolutionary theory, as first pronounced by Wallace and Darwin, is now taken for granted. How this multiplication proceeds is still controversial. That allopatric speciation, and particularly its special form of peripatric speciation (Mayr 1954, 1982c), is the most common mode is widely assumed. That speciation by polyploidy is common in plants is likewise accepted. How important other processes are, like sympatric and parapatric speciation, is still controversial.

Finally, the importance of natural selection, the theory that is usually meant by the modern biologist when speaking of Darwinism, is now firmly accepted by nearly everyone. Rival theories – like finalistic theories, neo-Lamarckism, and saltationism – have been so thoroughly refuted that they are no longer seriously discussed. Where the modern biologist perhaps differs from Darwin most is in assigning a far greater role to stochastic processes than did Darwin and the early neo-Darwinians. Chance plays a role not only during the first step of natural selection, the production of new, genetically unique individuals, but also during the probabilistic process of the determination of reproductive success of these individuals. Yet when one looks at all the modifications that have been made in the Darwinian theories between 1859 and 2004, one finds that none of these changes affects the basic structure of the Darwinian paradigm. There is no justification whatsoever for the claim that the Darwinian paradigm has been refuted and has to be replaced by something new. It strikes me as almost miraculous that Darwin in 1859 came so close to what would be considered valid 145 years later. And this extraordinary stability of the Darwinian paradigm justifies that it is so widely accepted as a legitimate foundation for a philosophy of biology and, in particular, as a basis of human ethics.

LITERATURE CITED

Darwin, C. 1859. *On the Origin of Species by Means of Natural Selection or the Preservation of Favored Races in the Struggle for Life.* London: John Murray [1964, Facsimile of the First Edition; Cambridge, MA: Harvard University Press].

Darwin, F. (ed.). 1887. *The Life and Letters of Charles Darwin, Including an Auto-biographical Chapter*, 2 vols. New York: Appleton.

Darwin, C. In *Charles Darwin's Natural Selection*, R. C. Stauffer (ed.). 1975. Cambridge: Cambridge University Press.

Eldredge, N., and S. J. Gould. 1972. Punctuated equilibria: an alternative to phyletic gradualism. In *Models in Paleobiology*, T. J. M. Schopf (ed.). San Francisco: Freeman, pp. 82–115.

Gould, S. J., and R. C. Lewontin. 1979. The spandrels of San Marco and the Panglossian paradigm: a critique of the adaptationist programme. *Proceedings of the Royal Society of London, Series B*, 205:581–589.

Herschel, J. F. W. 1861. *Physical Geography of the Globe.* London: Longmans, Green.

Hull, D. L. 1973. *Darwin and his Critics. The Reception of Darwin's Theory of Evolution by the Scientific Community.* Cambridge, MA: Harvard University Press.

Huxley, A. 1981. Anniversary address of the president. Supplement to Royal Society News, no. 12, pp. i–vii.

Jacob, F. 1977. Evolution and tinkering. *Science*, 196:1161–1166.

Kottler, M. J. 1978. Charles Darwin's biological species concept and theory of geographic speciation: the transmutation notebooks. *American Scientist*, 35:275–297.

Kotter, M. J. 1985. Charles Darwin and Alfred Russel Wallace. Two decades of debate over natural selection. In *The Darwinian Heritage*, D. Kohn (ed.). Princeton: Princeton University Press, pp. 367–432.

Lamarck, J.-B. 1809. *Philosophie zoologique, pour esposition des conserérations relatives de l'histoire naturelle des animaux*, 2 vols. Paris: Savy. [English translation by H. Elliot, The Zoological Philosophy, 1914, London: Macmillan.]

Margulis, L., and D. Sagan. 2002. *Acquiring Genomes: A Theory of the Origins of Species*. New York: Basic Books.

Mayr, E. 1954. Change of genetic environment and evolution. In *Evolution as a Process*, J. Huxley, A. C. Hardy, and E. B. Ford (eds.). London: Allen & Unwin, pp. 157–180.

Mayr, E. 1982a. Adaptation and selection. *Biologisches Zentralblatt*, 101:161–174.

Mayr, E. 1982b. *The Growth of Biological Thought*. Cambridge, MA: Harvard University Press.

Mayr, E. 1982c. Speciation and macroevolution. Evolution, 36:1119–1132.

Mayr, E. 1985. Darwin's five theories of evolution. In *The Darwinian Heritage*, D. Kohn (ed.). Princeton: Princeton University Press, pp. 755–772.

Mayr, E. 1992. Darwin's principle of divergence. *Journal of the History of Biology*, 25:343–359.

Monod, J. 1970. *Le Hasard et la Necessité*. Paris: Seuil.

Ospovat, D. 1981. *The Development of Darwin's Theory: Natural History, Natural Theology, and Natural Selection, 1838–1859*. Cambridge: Cambridge University Press.

Poulton, E. B. 1896. *Charles Darwin and the Theory of Natural Selection*. London: Cassell.

Sulloway, F. J. 1979. Geographic isolation in Darwin's thinking: the vicissitudes of a crucial idea. *Studies in the History of Biology*, 3:23–65.

Wright, S. 1967. Comments. In *Mathematical Challenges to the Neo-Darwinian Interpretation of Evolution*, P. S. Moorhead and M. M. Kaplan (eds.). Philadelphia: Wistar Institute Press, pp. 117–120.

7

Maturation of Darwinism

Even though Darwin had presented in 1859 the basic principles of Darwinism in great detail in *Origin of Species*, it took another eighty years for biologists to fully accept Darwinism. There were many reasons why there was so much disharmony in evolutionary biology during this long period. Perhaps the main reason was that the very concept of Darwinism continued to change over time. And different Darwinians endorsed different combinations of Darwin's five theories (see chapter 6). In my *One Long Argument* (1991), I describe nine different usages of the term Darwinism, more or less popular at different periods. Only a chronological treatment can do justice to the history of the concept of Darwinism.

Stages in the maturation of Darwinism

It is now obvious why there was such disharmony in evolutionary biology for the first eighty years. At the beginning, Darwinism simply meant anticreationism. An evolutionist was labeled a Darwinian as long as he adopted at least the theory that evolutionary change was due to natural causes and not to divine action (Mayr 1991, 1997). You were a Darwinian if you considered science a secular endeavor. Accordingly, opponents of natural selection such as T. H. Huxley and Charles Lyell were called Darwinians. No wonder the term Darwinism had so many meanings in the nineteenth century.

1859–1882

The first years after 1859 were a period of considerable confusion in evolutionary biology. To be sure, two of Darwin's five theories, an evolving world and common descent, were at once almost universally accepted. But his other three theories were not popular. Natural selection, in particular, was very much a minority view.

Transmutationism and the two transformationist theories were vastly more popular than Darwin's variational evolution (Mayr 2001). Darwin had given up on speciation and his vast efforts to explain the nature and the origin of variation had been unsuccessful. An inheritance of acquired characters was almost universally accepted. Darwin adopted it simultaneously with natural selection. It helped him to explain the ubiquity of new variation and for him it did not interfere with the primacy of natural selection. A large proportion of the naturalists accepted such a combination of natural selection and inheritance of acquired characters (Plate 1913).

1883–1899

In 1883 August Weismann, the greatest evolutionist after Darwin, published his refutation of an inheritance of acquired characters, and he was followed in this by Alfred Russel Wallace and other Darwinians. Through his general theorizing Weismann prepared the way for Mendel's discovery, as Correns has rightly pointed out. Romanes (1894) coined the term *neo-Darwinism* for this new kind of Darwinism without an inheritance of acquired characters. Some recent historians have misused the term neo-Darwinism for the compound of theories emerging from the evolutionary synthesis, but this is not correct. Neo-Darwinism is the designation for Weismann's revised Darwinism (excluding any inheritance of acquired characters).

1900–1909

When Gregor Mendel's work was rediscovered in 1900, many hoped the new science of genetics, with its laws of inheritance, would provide the answers to the great controversies over evolution that had raged since Darwin's day.

The leading Mendelian geneticists most interested in evolution – Hugo de Vries (one of Mendel's "rediscoverers"), William Bateson, and Wilhelm Johannsen – however, unfortunately rejected natural selection, the keystone of Darwin's thought. De Vries certainly adopted instead saltationism. According to him, a new species originates by a major genetic mutation that in a single jump (saltation) gives rise to a new species. This theory of saltations dominated evolutionary genetics from 1900 to about 1915. Unfortunately, because this Mendelian-mutationist theory of evolutionary change was widely accepted by geneticists, it was considered by most naturalists under the name Mendelism to be the genetic theory of evolution, even though it was not held by some "Mendelians." Because the naturalists on the whole believed in gradual evolution and in populational variation, they found saltationist Mendelism quite unacceptable and this created a seemingly unbridgeable gap among the evolutionists.

Since Darwin's time and even before, those who had observed living populations realized that the origin of new species was usually a gradual process. These naturalists would have nothing to do with mutationism and instead held fast to the concept of gradual evolution, which had been first articulated by Jean Baptiste Lamarck in the early nineteenth century. Because gradualism is explained by Lamarckian transformationism, these naturalists became Lamarckians.

However, the Mendelians were not the only contemporary geneticists. There were others, including Nilsson-Ehle, Baur, Castle, East, and, in Russia, Chetverikov, who accepted the occurrence of small mutations and of natural selection. The existence of these gradualist geneticists, however, was ignored by the naturalists, who concentrated their attack on the saltationism of de Vries and his followers. On the whole, the Darwinian interpretation, with its emphasis on the role of natural selection, had a nadir of popularity in this early period of genetics (early 1900s) and was, at that time, often declared to be dead.

1910–1932

The new science of genetics developed new methodologies and a new theoretical framework, moving it away from the transmutationism of the Mendelians. Starting around 1910 in the laboratory of T. H. Morgan at Columbia University, New York, a new generation of geneticists arrived on the scene whose findings contradicted the views of the early Mendelians. In their experiments with fruit flies (*Drosophila*) these researchers discovered that most mutations are small enough to permit a gradual change in populations; no sudden jumps were required. Soon the saltationism of the Mendelians was considered obsolete. Between 1915 and 1932, the mathematical population geneticists, Fisher (1930), Wright (1931), and Haldane (1932), showed that genes with only small selective advantages in due time could be incorporated into the genotype of populations. Phyletic evolution could now be explained in terms of the new genetics. Unfortunately, most naturalists were unaware of these developments and were still fighting the antigradualism of the early Mendelians.

According to the more-or-less unified theory of Fisher and his colleagues, evolution was defined as a change in gene frequencies in populations, a change brought about through the gradual natural selection of small random mutations. By 1932, a consensus was reached on these findings among the various feuding schools of geneticists. It was a synthesis between mathematical population geneticists and Darwinian selectionists. This synthesis, which one might call the Fisherian synthesis after its greatest representative, solved one of the two major problems of evolutionary biology, the problem of adaptedness. Adaptedness is indeed – as Darwin had believed – the result of natural selection acting on abundant variation. Unfortunately, this Fisherian synthesis of the late 1920s has been confused by many historians with a second synthesis involving biodiversity.

The explanation of the origin of biodiversity

Adaptation is only one half of the story of evolution. Evolutionary biology is concerned with two discrete processes: phyletic evolution over time within a given population and the origin and multiplication of species. Fisher, Wright, and Haldane were primarily interested in determining how a population evolves as the environment changes. This branch of evolutionary biology has been referred to as the study of *anagenesis*. The naturalists, by contrast, were more interested in diversity and in determining how new species branch off from their parent species. This study of the origin of biodiversity is often referred to as *cladogenesis*. In other words, the mathematical population geneticists were concerned with the vertical or "time" dimension of evolution (changes over time within a given population), whereas the naturalists were mostly concerned with the horizontal or geographic dimension of evolution (the production of new species at a given time).

This second great evolutionary problem – the multiplication of species, or the origin of biodiversity – remained unsolved by the Fisherian synthesis. The geneticists were unable to explain speciation because their methods restricted them at any one time to the study of a single population, a single gene pool. While the Fisherian synthesis resolved the conflict between Mendelian genetics and natural selection, it failed to address the conflict between mathematical genetics and biodiversity. Fisher, Haldane, and Wright were aware of the problem of the origin of biodiversity and vaguely referred to it, particularly Wright, but they did not seem to realize the role played by the geographic location of populations and by isolation. Providing an explanation of how life proliferates into so many diverse forms at a given time – as opposed to just one form that continually changes over time – was the achievement of a second synthesis, initiated in 1937 by Dobzhansky's work *Genetics and the Origin of Species*.

Actually, the European naturalists, through their work in taxonomy and natural history, already had an explanation for the origin of biodiversity in the 1920s. According to these naturalist-taxonomists, speciation occurs when two populations of a species become physically separated from each other and during this spatial isolation become reproductively isolated, either through the development of sterility barriers or of behavioral incompatibilities (isolating mechanisms). Sometimes the geographic separation occurs because of a new physical barrier (a new mountain range or arm of the sea) (dichopatric speciation), and sometimes it occurs because a founder population establishes itself beyond the species' previous range (peripatric speciation). If the geographically isolated population contains the potential for a major divergence, a new species will branch off from the parent species. Both dichopatric and peripatric speciation are referred to as geographic speciation.

These ideas about speciation remained unknown to the laboratory geneticists. Meanwhile, the naturalists were likewise unable to arrive at a full understanding of evolution because of their ignorance of the recent developments in genetics. They were still arguing against the saltational model of the early Mendelians, which Fisher and his colleagues had long since refuted. Because naturalists like Stresemann and Rensch and French zoologists could not accept de Vries's large genetic jumps and were unaware of the small mutations discovered by later geneticists, they turned to Lamarckism to explain gradual evolution. The naturalists (myself included) accepted the Lamarckian notion that variation arose through the use or disuse of existing body parts and that these acquired "characters" could be passed on to offspring. Even though most naturalists staunchly defended selection against the errors of the saltationists, the naturalists also retained an outdated Lamarckian explanation of variation. Thus, in spite of great advances made in both genetics and taxonomy, there was a deep chasm of misunderstanding between the experimental geneticists and the naturalists-taxonomists. Some mathematical geneticists credit Wright's "landscape model" with a contribution to the solution of the geographic speciation theory, but a critical analysis of the relevant theory gives no support to this claim.

This chasm was finally bridged by the "evolutionary synthesis" of the 1940s. As mentioned, there was an earlier synthesis, that between genetics and Darwinism. I referred to it as the Fisherian synthesis (origin of mathematical population genetics). This has often been confused with the later Dobzhansky synthesis (origin of biodiversity) in histories written by geneticists. This earlier ("Fisherian") synthesis dealt with single gene pools, with single populations, with genetic variations, and with the origin of adaptation. It made no contribution to the solution of the problem of biodiversity. The chasm between the geneticists, particularly interested in variation and adaptation, and the taxonomists, interested in the origin of biodiversity, remained.

Actually, there was no longer any conflict between the micromutational interpretation of evolution (as now held by the geneticists) and the evolutionary ideas of the naturalists. However, the geneticists dealt only with a given population and the field of the origin of biodiversity was beyond their methodology. Hence, there was still a considerable gap between the geneticists and the naturalists (taxonomists). The bridging of this gap was initiated in 1937 by the publication of Dobzhansky's *Genetics and the Origin of Species*. Dobzhansky, by his background, was ideally qualified for this task. A naturalist from boyhood on, he had received his biological education in Russia, where he was interested in individual and geographic variation and in speciation in a group of beetles (Coccinellidae). At age 27 he came to America and joined T. H. Morgan's laboratory where he became thoroughly acquainted with modern evolutionary genetics. The happy result of these two very different influences was his book *Genetics and the Origin of Species*, published in 1937. This book showed the geneticists and the naturalists that their theories of evolution were perfectly compatible and that it was possible to have a synthesis of the two major areas in evolutionary biology, the study of phyletic evolution in populations (anagenesis) and the origin of biodiversity (species, speciation, macroevolution) (cladogenesis). This synthesis of the two fields was completed in follow-up publications by Mayr, *Systematics and the Origin of Species* (1942); Huxley, *Evolution, the Modern Synthesis* (1942); Simpson, *Tempo and Mode in Evolution* (1944); and Stebbins, *Variation and Evolution in Plants* (1950); and on the European continent by B. Rensch (1947).

This synthesis of the 1940s dealt primarily with the origin and the significance of biodiversity: how and why new species arise. Every population has to be well adapted at all times and this accounts for the changes that occur in a species over time. But a population does not have to produce new species to remain adapted. The mechanisms that produce new species require very different explanations from the mechanisms studied by the geneticists that maintain adaptedness.

Another major achievement of the evolutionary synthesis was to establish a common front of the true Darwinians against the three non-Darwinian theories of evolution that were still widely held around 1930: Lamarckism (still accepted by many naturalists), saltationism [promoted by Schindewolf (1950) and by Goldschmidt (1940) with his "hopeful monsters"], and orthogenesis (a belief in some sort of goal-directed, teleological component in evolution). After the synthesis, these three theories no longer played a role in serious evolutionary discussions (Mayr 2001).

When in 1947 the evolutionists met in Princeton at a symposium to celebrate the synthesis they found that indeed a consensus had been largely reached and that the great controversies of the preceding fifty years were now a matter of history. Only one serious disagreement between the two camps was left, relating to the object of selection. This, for the naturalists, as it had been for Darwin, was the individual, while for the geneticists it was the gene, in part for the sake of ease of computation. Actually it is a rather important difference because it illustrates the reductionist tendency of the population geneticists, while the leading architects of the synthesis, particularly the naturalists, were strongly holistic in their views. Even before the synthesis, I, like most naturalists, was a holist. Evolution for me concerned the whole organism, and the organism as a whole was the target of selection. This was, of course, the Darwinian tradition. I admit that during the synthesis, I used the standard formula of the geneticists that "evolution is a change in gene frequencies," even though it was actually incompatible with my holistic thinking. But I did not appreciate this contradiction until many years later (Mayr 1977). Actually, in spite of the synthesis, the definition of evolution (whether reductionist or holist) continued to be the major point of disagreement between the geneticists and the naturalists. For the naturalists, evolution is more than a change in gene frequencies; it is the acquisition and maintenance of adaptedness and the origin of new biodiversity.

Immediately after the synthesis came the molecular revolution, a truly revolutionary episode in the history of biology. Avery showed in 1944 that the genetic material consist not of proteins but of nucleic acids; Watson and Crick discovered in 1953 the structure of DNA, which permitted explaining the activities of the DNA. Their discovery provided a critical new dimension for genetic analysis and solved innumerable previously insoluble questions. Finally Jacob and Monod showed in 1960 that there are various kinds of DNA and, in particular, a special regulatory DNA that controls the activity of the structural genes. These discoveries produced such a cataclysmic change of prevailing ideas that one was justified to expect a drastic effect on Darwinism.

Indeed, molecular biology made countless important contributions to our understanding of evolution. It showed that the genetic code is essentially the same from the primitive bacteria to the multicellular higher organisms. This proves that all life that now exists on earth descended from a single origin. Molecular biology also showed that information can be transferred only from nucleic acids to proteins and not from proteins to nucleic acids. This is the reason why there cannot be any inheritance of acquired characters.

Genomics

The greatest impact of the molecular revolution on evolutionary biology was coming from genomics, the comparative study of gene sequences. It showed that many genes are very old. For instance, some mammalian genes can be identified with genes in nonchordate phyla and even with prokaryote genes. Genomics permits the study of the effect of the replacement of single base pairs, the effect of the insertion of noncoding DNA, the shift of genes by lateral transfer, and the effect of all the numerous changes of genes and of their position on chromosomes. The invention of the molecular clock by Zuckerkandl and Pauling has been an enormous contribution to evolutionary methodology. Genomics is in the process of developing into a major branch of evolutionary genetics. It cannot be dealt with in a few words and I must refer to the relevant literature (Campbell and Heyer 2002).

The molecular revolution is particularly important for two reasons. It led to a revival of a number of divisions of classical biology, such as developmental biology and all aspects of gene physiology that had been neglected earlier in the century. By adopting molecular methods and theories, these areas experienced a revitalization and an approach to the modern branches of biology. Perhaps the most interesting other development was that through molecular biology numerous physicists and biochemists became interested in evolution. This resulted in much active bridge-building among branches of biology that previously had little understanding of each other. Thus, molecular biology made a major contribution to the unification of biology that took place in the twentieth century. A study of almost any current issue of *Evolution*, *The American Naturalist*, or other evolutionary journals shows how greatly molecular methods have contributed to the solution of evolutionary problems.

However, the gene-centered approach of most molecular biologists has led to some disagreements. For instance so-called neutral evolution is considered by many molecular biologists an important mode of evolution, but it is ignored by naturalists, because neutral genes are not visible in the phenotype.

The sequence of base pairs in the genome provides an enormous amount of information on the relationship and phylogeny of organisms. The standard morphological characters, used since the beginning of phylogenetic studies, were sometimes insufficient to provide a reliable phylogeny. The methods of molecular biology have supplied abundant information to permit a revolutionary restructuring of the phylogeny of many groups of organisms.

The robustness of the current Darwinian paradigm

The period from the 1940s (evolutionary synthesis) to the present has been a period of great advances in biology, including the origin and spectacular rise of molecular biology. One might have expected that it would have necessitated a thorough revision of Darwinism. Unexpectedly, nothing of the sort happened. The Darwinian paradigm produced in the 1940s during the evolutionary synthesis was able to resist without any major revision all the attacks against it during the last fifty years. This suggests that one might cautiously believe that the Darwinian paradigm that was adopted during the evolutionary synthesis is essentially valid. The basic Darwinian formula – evolution is the result of genetic variation and its ordering through elimination and selection – is sufficiently comprehensive to cope with all natural eventualities. To search for a new evolutionary theory (paradigm) seems now a futile enterprise. For the last fifty years or longer almost every year a new paper or even book was published in which a serious error or omission in Darwinism was claimed. The author proposed a new theory, or new theories, which he or she claimed would correct this error and fill the gap. Alas, not a single one of these proposals turned out to be constructive. Invariably the now classical Darwinism was confirmed and it was possible to refute the putative improvements or corrections. This suggests to me that Darwinism is nearing full maturity. There are, of course, still numerous unsolved puzzles like the function of much of the noncoding DNA, but I do not see how the solution of any of the remaining puzzles can have any noticeable effect on the basic Darwinian paradigm.

The major controversies in evolutionary biology in recent years, such as the importance of adaptation, the role of chance, population thinking, the gradualness of evolution, the steadiness of evolutionary rates, etc., deal with individuals and populations, not with genes. Even the discovery by Jacob and Monod that there are different kinds of genes, structural and regulatory, did not affect Darwinian theory. The two main reasons for the robustness of the Darwinian paradigm are probably the failure of reductionism and the simplicity of the basic Darwinism.

The Darwinism of the present

This is a highly abridged account of the history of Darwinism since 1859 and particularly since the 1920s. I have recently published several more detailed narratives of the history of the synthesis, with a discussion of various errors and inaccuracies that mar the accounts of some geneticists and historians (1992, 1993, 1997, 1999a, 1999b, 2001). In particular, I point out that some historians confuse the Fisherian synthesis of the 1920s with the synthesis of the 1940s.

What name should we apply to the version of Darwinism developed in the 1940s? Erroneously, it has often been referred to as *neo-Darwinism*. But this choice is clearly wrong. Neo-Darwinism is the term given by Romanes in 1894 to the Darwinian paradigm but without soft inheritance (i.e., without a belief in an inheritance of acquired characters), but this has been true for all Darwinism since the 1920s. The new evolutionary theory, the product of the synthesis of the theories of the students of anagenesis and of cladogenesis, has been called the synthetic theory of evolution. Actually the best solution would be to call it again simply Darwinism. Indeed it is essentially Darwin's original theory with a valid theory of speciation and without soft inheritance. But because such inheritance was refuted more than 100 years ago, no mistake can occur if we go back to the simple term Darwinism, because it encompasses the essentials of Darwin's original concept. In particular it refers to the interplay of variation and selection, the gist of Darwin's paradigm. It certifies that the evolutionary paradigm adopted by the modern evolutionists, after a long period of maturation, is best simply referred to as Darwinism.

LITERATURE CITED

Campbell, A. M., and L. J. Heyer. 2002. *Discovering Genomics, Proteomics, and Bioinformatics*. San Francisco: Benjamin Cummings.

Dobzhansky, Th. 1937. *Genetics and the Origin of Species*. New York: Columbia University Press.

Fisher, R. A. 1930. *The Genetical Theory of Natural Selection*. Oxford: Clarendon Press.

Goldschmidt, R. 1940. *The Material Basis of Evolution*. New Haven: Yale University Press.

Gould, S. J. 1977. The return of hopeful monsters. *Natural History*, 86 (June/July):22–30.

Haldane, J. B. S. 1932. *The Causes of Evolution*. New York: Longman, Green.

Huxley, J. 1942. *Evolution. The Modern Synthesis*. London: Allen & Unwin.

Mayr, E. 1942. *Systematics and the Origin of Species*. New York: Columbia University Press.

Mayr, E. 1977. The study of evolution, historically viewed. In *The Changing Scene in Natural Science, 1776–1976*, C. E. Goulden (ed.). Philadelphia: Academy of Natural Sciences. Special Publication 12, pp. 39–58.

Mayr, E. 1991. *One Long Argument*. Cambridge, MA: Harvard University Press.

Mayr, E. 1992. Controversies in Retrospect. In *Oxford Surveys in Evolutionary Biology*, vol. 8, D. Futuyma and J. Antonovics (eds.). Oxford: Oxford University Press, pp. 1–34.

Mayr, E. 1993. What was the evolutionary synthesis? *Trends in Ecology and Evolution* 8:31–34.

Mayr, E. 1995. Darwin's Impact on Modern Thought. Proceedings of the American Philosophical Society, 139:317–325.

Mayr, E. 1997. The establishment of evolutionary biology as a discrete biological discipline. *BioEssays* 13:263–266.

Mayr, E. 1998. New preface to *The Evolutionary Synthesis: Perspectives on the Unification of Biology*, E. Mayr and W. B. Provine (eds.). Cambridge, MA: Harvard University Press, 1998, pp. ix–xiv.

Mayr, E. 1999a. Postscript: Understanding evolution. *Trends in Ecology and Evolution*, 14(9):372–373. [Written for the receipt of the Crafoord Prize.]

Mayr, E. 1999b. An Evolutionist's Perspective. *Quarterly Review of Biology* 74:401–403.

Mayr, E. 1999c. Foreword to *Systematics and the Origin of Species* (new edition). Cambridge, MA: Harvard University Press, pp. xxiii–xxxv.

Mayr, E. 2001. *What Evolution Is*. New York: Basic Books.

Plate, L. 1913. *Selektionsprinzip und Probleme der Artbildung; ein Handbuch des Darwinismus*, 4th ed. Leipzig: W. Engelmann.

Plate, L. 1923. *Die Abstammungslehre*. Jena: Gustav Fisher.

Rensch, B. 1947. *Neuere Probleme de Abstammungslehre*. Jena: Enke.

Romanes, G. J. 1894 [1892–1897]. *Darwin, and After Darwin: An Exposition of the Darwinian Theory and a Discussion of Post-Darwinian Questions*, vols. 1–3. Chicago: Open Court Publishing.

Schindewolf, O. H. 1950. *Grundfragen der Paläontologie*. Stuttgart: Schweizerbart.

Simpson, G. G. 1944. *Tempo and Mode in Evolution*. New York: Columbia University Press.

Stebbins, G. L. 1950. *Variation and Evolution in Plants*. New York: Columbia University Press.

Woese, C. R. 2002. On the evolution of cells. *Proceedings of the National Academy of Sciences*, 99:8742–8747.

Wright, S. 1931. Evolution in Mendelian populations. *Genetics*, 16:97–159.

8

Selection

THE CORNERSTONE OF DARWIN'S PARADIGM OF EVOLUTION was the theory of natural selection. Yet, of all his theories this was the last one to be adopted by his followers. It took some eighty years before it was fully accepted by biologists and, of course, even today it still encounters a good deal of resistance among laypersons, particularly those with religious commitments. Actually at the beginning there were good reasons for resistance. Most importantly, for a long time there was little convincing evidence for the occurrence of selection in nature. Such evidence has now been provided abundantly, both in the field and in the laboratory (Endler 1986, Futuyma 1999). But there was also considerable uncertainty about various specific aspects of the selection process.

I will not present a full treatment of the subject natural selection in this chapter for I have done so quite recently in *What Evolution Is* (2001:

chapter 6, pp. 115–146). Instead I will single out for special treatment various aspects of selection about which there are still uncertainties.

What is selection?

In view of the persistent controversies, from 1859 on, concerning the nature of selection, it would seem most helpful to begin with a concise definition of selection, but this cannot be done owing to the arguments on the nature of this process. In 1963 I defined natural selection as nonrandom "differential reproductive success." And this is even today a valid formulation, but it stresses the outcome of this process rather than its mechanism.

For Darwin and most of his followers for the next sixty years, natural selection was a rather simple process. Owing to the struggle for existence, there was enormous mortality in every generation and only the best survived. Fortunately, nature offered a virtually inexhaustible supply of variation and through the survival of the best there was steady evolutionary advance.

Darwin borrowed the term selection from the vocabulary of the animal breeders and plant cultivators. But he overlooked that the breeders actually utilized two very different approaches to improve their stocks and so does nature. According to one of these approaches, those individuals are selected as breeding stock for the next generation that had special characteristics that represented the ideal of what the breeders aimed for in their selection. They would simply say that they would choose "the best individuals" of their flocks as their breeding stock. It was this method that Darwin apparently had in mind when he used the word "select."

However, the breeders often used instead a different method to which they referred as "culling." In this method only the truly inferior individuals were eliminated and all the remaining individuals were used for breeding. This, of course, was not at all a "selection of the best." Nature uses the same two methods. In a harsh year as far as survival factors are concerned, only the best individuals survive; all others are eliminated. In a mild year only the worst are culled and most individuals survive. At the beginning of the next breeding season, as a result of such great survival a much more diversified population is available for the action of sexual selection and for selection contingencies. The existence of this culling method was soon pointed out by Herbert Spencer when he called natural selection a "survival of the fittest." He should have said "survival of the fitter." The survivors are those left over after all the inferior individuals have been eliminated. This elimination process is not at all a "selection of the best."

Curiously, it has never been remarked that the consequences of an elimination process may be quite different from those of a selection process. A selection process results in the survival of the truly best and there will be only relatively few individuals that qualify for such a designation. In a process of actual selection a bird with a cumbersome tail such as that of a peacock would never emerge as "the best." By contrast, elimination would leave in an average year a much larger percentage of survivors than would the selection of only the best. This large pool of survivors provides ample material for sexual selection and for chance. It provides an explanation for the haphazardness of much of evolutionary change. Evolution by elimination provides a far better explanation for the actual course of events during evolution than the "selection of the best" of the classical evolutionary literature. The elimination of the inferior takes place, of course, simultaneously with the selection of the best, but it differs in strength in different situations. The unpredictability of much of evolution, described so graphically by Gould in his *Wonderful Life* (1989), is well explained by the process of elimination but could not be accounted for by a restriction to the selection of the best.

Actually a selection of the best and an elimination of the worst take place simultaneously. The two processes also can be conceived of as occurring in parallel. Furthermore, natural selection is now seen as two very different processes, natural selection proper (survival selection) and differential production of offspring due to variation in the ability to deal with environmental factors other than mates and sexual selection (selection for reproductive success) – specifically, success in competition for mates. In certain kinds of organisms, such selection for reproductive success may be more important than survival selection.

Natural selection, a two-step process

An apt summary of Darwin's paradigm is the phrase "variation and selection." However, almost all the way back to the publication of the *Origin* (1859) there has been a controversy over whether variation or selection is more important. For some Darwinians, every component of the phenotype was the direct result of selection. For others, many of the aspects of the phenotype were chance phenomena that happened during evolution. We now realize that this argument is largely misleading. Every step in the evolutionary process is affected both by variation and by selection.

One achieves a better-balanced view if one considers selection a two-step process. Every population in every generation must go through both of these steps. The first step is the production of variation. Every potential object of selection goes through several processes: mutation, the restructuring of chromosomes during meiosis, the random movement of chromosomes to different daughter cells during the reduction division, and the chance aspects of the meeting of the two gametes. At this first step, everything is chance, everything is randomness. The second step of selection is the fate of the new zygote from its formation to its successful reproduction. At this step, selection is the dominant factor even though chance still plays a considerable role.

Uncertainty still exists on the question of how much variation is available in each generation. Under the classical view, which assumes that after a bout of natural selection only the best remain, relatively little variation is available. According to the elimination model, however, where only the truly inferior individuals are vulnerable, a great deal of variation still remains among the less inferior individuals. This is rarely emphasized in arguments over natural selection. It means that in spite of elimination there will always be abundant variation, consisting of not only the very best individuals but indeed of all that are not so bad that they have to be removed by elimination. To call attention to the occurrence of two largely independent processes – the elimination of the worst and the selection of the best – I have described them as occurring sequentially. In reality, the two processes take place simultaneously. What is important is the size of the generation between the two extremes, the best and the worst. The larger this generation is and the richer in variation, the greater the opportunity will be for mate selection ("sexual selection") and for contingencies. No peacock with its cumbersome tail could have evolved if always only those peacocks had been selected that were "best" for survival. However, a process of mild elimination made a large reservoir of variation available. This provided the next generation with a much richer supply of variants than "selection of the best." The opponents of Darwinism always questioned how selection could have tolerated so many rather aberrant evolutionary trends. Indeed, they would not have had a chance under the "selection of the best" principle. However, under the more tolerant concept of elimination a wide range of phenotypes variable in their ability to deal with environmental contingencies is still available (as long as they are not so bad that they have to be removed by elimination). In such a situation, chance would be particularly important in deciding which would be the lucky ones to produce the next generation. Let us remember also that phenotypes are the object of selection, not single genes.

Some enthusiasts have claimed that natural selection could do anything. This is not true. The options of selection are quite limited. Darwin was not right when he claimed "natural selection is daily and hourly scrutinizing throughout the world, every variation even the slightest" (1859: 84). Actually, selection is quite liberal in what it makes available to sexual selection and to chance. Thus, some even rather aberrant individuals may become the progenitors of new evolutionary lineages. This is a point Gould (1989) has rightly stressed. Furthermore, the maintenance of perfect adaptedness of a population may be prevented by numerous constraints (see Mayr 2001: 140–143). It must be realized at all times that natural selection is a population phenomenon dealing with individuals. Any typological interpretation is in error.

Selection for reproductive success

When we speak of natural selection, unconsciously we always think of the struggle for existence. We think of factors that favor survival such as a capacity to overcome adverse weather conditions, to escape enemies, to better cope with parasites and pathogens, and to be successful in competition for food and habitation – in short, to have any property that would enhance the chances for survival. This "survival selection" is what most people have in mind when they speak of natural selection. Darwin, however, saw quite clearly that there were other factors enhancing the probability of leaving offspring. Any such factors can be referred to as selection for reproductive success involving competition for mates. Among these factors, Darwin singled out for special attention are traits that affect success in competition for mates either via male – male combat or by female choice. He combined these two modes of mate selection under the name *sexual selection*. To indicate how important Darwin considered this process, he devoted to it two-thirds of *The Descent of Man* (1871).

It has become clear since Darwin's day that sexual selection is only one of a far wider realm of phenomena and that instead of sexual selection these activities are better referred to as "selection for reproductive success in direct competition with conspecifics." It includes also such phenomena as parent–offspring conflict, sibling rivalry, unequal parental investment, unequal rates of division in prokaryotes, and most of the phenomena studied by sociobiology. Unlike survival selection, genuine selection is involved in any kind of selection for reproductive success. Recent studies (Carson 2002) have shown that female choice may be important in *Drosophila* even in ordinary mate selection and presumably also in many other species. Considering how many new kinds of selection for reproductive success are discovered year after year, I am beginning to wonder whether this process is not even more important than survival selection, at least in certain higher organisms. Curiously, factors that contribute to reproductive success were largely neglected by evolutionists until around 1970. At that time naturalists rediscovered Darwin's important finding (1871) that females may play a decisive role in the choice of their mates. This favors the evolution of male characteristics attractive to females.

Furthermore, particularly as a result of the study of social hymenoptera (ants, bees), it was shown how many life history factors favor the selection for reproductive success. And this led to the development of a flourishing branch of biology, *sociobiology* (Wilson 1975). As J. B. S. Haldane was the first to point out, altruistic behavior toward close relatives will be favored by this kind of selection (*kin selection*), and this explains the existence of neutral castes among the social hymenopterans (Hamilton 1964).

Levels of selection

One of the most basic questions of evolutionary biology is what objects are selected in the process of natural selection. Lloyd (1992) found nearly two hundred references to books and papers by biologists and philosophers, beginning with Darwin, that treated this question, "and these represent just a fraction of the literature on the topic," she reports. Indeed in the recent literature the answer to this question has been argued each year by at least a half dozen authors. [This analysis is not a review paper. The listing of the literature is therefore reduced to a minimum. Other relevant titles can be found in the works of Lloyd (1992) and Brandon (1990).] An analysis of this literature has convinced me that the major causes of controversy on this subject are some basic conceptual differences, as well as the opponents' failure to adhere to a rigorous definition of the terms. Evidently, an approach that attempts a careful critique of the arguments of the opposing parties is needed. This is what I am attempting here.

The objects of selection

The difficulty begins with the exact description of the process of selection. After Darwin had discovered his new principle, he searched for an appropriate terminology and thought he had found it in the term selection (1859), a term animal breeders used for the choice of their breeding stock. However, as first Herbert Spencer and then Alfred Russel Wallace pointed out to him, there is no agent in nature that, like the breeders, "selects the best." Rather, the beneficiaries of selection are all the individuals left over after the less-fit individuals have been eliminated. Natural selection thus is a process of "nonrandom elimination." Spencer's statement, "survival of the fittest," was quite legitimate, provided the term fittest is properly defined (Mayr 1963:199) as reproductive success.

Even though most evolutionists now agree that the individual organism is the principal object of selection, there is still considerable discussion about the validity of accepting additional objects of selection. Some years ago, I attempted to list all the terms used by various authors (Mayr 1997); here I present a revised version of this listing.

The gene

The rediscovery of Mendel's work in 1900 led geneticists increasingly to replace the individual for mathematical tractability by the gene as the object of selection. By 1930 this was the standard viewpoint among the geneticists, particularly the mathematical population geneticists. It was at this period that a definition became popular, "evolution is the change of gene frequencies in populations." For the naturalists, however, the object of selection continued to be the individual. But even among the geneticists, doubts began to be expressed in the 1940s and 1950s. A group of holistic population geneticists, including Lerner (1954), Mather (1943), and Wallace et al. (1953) began to stress the cohesion of the genotype. I myself attacked the reductionism of "bean bag genetics" (Mayr 1959) and stated bluntly that the phenotype was the target of selection (Mayr 1963:279–296). But the replacement of the gene by the individual among the geneticists was a slow process.

The idea of the gene as the target of selection was still widely accepted as late as 1970 – for instance, by Lewontin. But eventually it was severely criticized (Wimsatt 1980, Sober and Lewontin 1982) with the critics pointing out that "naked genes," "not being independent objects" (Mayr 1976), are not "visible" to selection and therefore can never serve as a target. Furthermore, the same gene – for instance, the human sickle cell gene – may be beneficial in heterozygous condition (in *Plasmodium falciparum* areas) but deleterious and often lethal in the homozygous state. Many genes have different fitness values when placed into different genotypes. Genic selectionism is also invalidated by the pleiotropy of many genes and the interaction of genes controlling polygenic components of the phenotype. On one occasion Dawkins (1982: point 7) himself admits that the gene is not an object of selection: "genetic replicators are selected not directly, but by proxy . . . [by] their phenotypic effects." Precisely! Nor are combinations of genes – as, for instance, chromosomes – independent objects of selection; only their carriers are.

The gamete

Because only a small fraction of all eggs are fertilized and only an infinitesimal fraction of male gametes succeed in fertilizing an egg, gametes are potentially a category of entities subject to intense selection. But it is difficult to measure the fitness of gametes. Gametes have two sets of characteristics. One consists of the attributes a gamete has to have to facilitate fertilization. Evidently, the ability to swim rapidly, to be able to sense unfertilized eggs, and to be able to penetrate the egg membrane are properties of the spermatozoa that are most important in achieving success. Much experimental work on these properties has been carried out in recent years. However, these phenotypic properties of the spermatozoa presumably are produced by the paternal testis and are probably part of the extended phenotype of the male parent. They have nothing to do with the haploid genome of the gametes, which, so far as we can tell, has no influence whatsoever on the fertilizing capacity of these gametes. In some organisms, gametes (e.g., plant pollen grains and free-swimming gametes in aquatic organisms) seem to have gamete-specific properties influencing mating success. In some organisms, the reproduction tract of the female has an important influence on the fate of ejaculates (Eberhard 1996).

The individual organism

From Darwin to the present day most evolutionists (Lloyd 1992) have considered the individual organism to be the principal object of selection. Actually, the phenotype is the part of the individual that is "visible" to selection (Mayr 1963: 184, 189). Every genotype, interacting with the environment, produces a range of phenotypes, called by Woltereck (1909) the "norm of reaction." Therefore, when an evolutionist says that the "genome is a program that directs development," it would be wrong to think of it in a deterministic way. The development of the phenotype involves many stochastic and environmental processes, which preclude a one-to-one relation between genotype and phenotype. This is, of course, precisely the reason why we must accept the phenotype as the object of selection rather than the genotype.

Different phenotypic expressions of the same genotype may differ considerably in their fitness value. What is visible to selection is the phenotype that "screens off" the underlying genotype (Brandon 1990). The term phenotype refers not only to structural characteristics but also to behavioral ones and to the products of such behavior such as bird nests and spider webs. Dawkins (1982: point 7) has introduced for such characteristics the very useful term the *extended phenotype*. However, such species-specific behaviors are programmed in the neural system of these individuals and thus do not differ in principle from the morphological aspects of the phenotype.

In this account, when I refer to the term individual, I always mean what the word individual means in the daily language – that is, the individual organism. Philosophers have also applied the term individual to "particulars," like the species. I have avoided this designation because it is apt to create confusion.

The object of selection, individual or gene?

Genetics did not exist in 1859 when Darwin published the *Origin of Species*. For him, the individual was obviously the object of selection. And so it was for most Darwinians until the rise of genetics. Then most geneticists adopted the gene as the target of selection, while for most taxonomists and naturalists it remained the individual. During the evolutionary synthesis in the 1940s the two groups achieved wide-reaching consensus, but curiously one major difference remained: most geneticists still considered the gene the object of selection, while for the naturalists it remained the individual. However, in the 1960s and 1970s an ever-increasing number of geneticists realized that the isolated gene is not visible to selection and that the formula "evolution is a change of gene frequencies" is quite misleading (Mayr 1977). By the 1980s most geneticists had completed the shift (Sober 1984) and most evolutionists had learned that one must distinguish the two questions, *selection of?* and *selection for?* (see below).

When Williams (1966) rejected group selection, he could have chosen instead either the individual or the gene as the object of selection. Even though by that time perhaps the majority of the evolutionists had returned to Darwin's choice of the individual, Williams chose the gene ("alternative alleles in Mendelian populations") (p. 3). He was not unaware of the significance of the individual: "We can surely say that individuals characterized by fleetness, disease resistance, sensory acuity, and fertility are more fit than those that are less fleet, less resistant, etc." (p. 102), but he also says, "We cannot measure fitness by evolutionary success on an individual basis" (p. 102).

William's choice of the gene as the principal object of selection was adopted by a number of evolutionists, most enthusiastically by Dawkins, particularly in his *The Selfish Gene* (1976). Yet, except for Dawkins and a few of his followers, the rejection by geneticists of the gene as the object of selection was by then essentially complete.

Evidently a major reason for Williams's choice of the gene rather than the individual was the stability of the gene. He insists that "only the gene is stable enough to be effectively selected" while "genotypes have limited lives and fail to reproduce themselves" (p. 109). He evidently failed to realize that the frequency of a gene in a population can steadily increase no matter how many recombinations it is subjected to in various genotypes in the course of succeeding generations. Mendel's principle of particulate inheritance permits a gene to be unaffected by recombination. There is no blending inheritance.

Kin selection

Haldane was the first evolutionist to point out that selection of relatives who share part of your genotype would be of selective significance. This kind of selection is called *kin selection*. This is, of course, obvious in parent (mother)–offspring relationships, but, as Haldane emphasized, it is in principle also true for more distant relatives. In this case, kin selection and social group selection overlap and are difficult to discriminate from each other. Most kin selection is simultaneously also social-group selection. Members of both groups know each other from birth on and are accustomed to reciprocal helpfulness. There is no way to partition it into a kin selection and a social group selection component. For other problems with kin selection, see Mayr (2001:132, 257).

Group selection

There has been a long and bitter controversy about whether groups as cohesive wholes can serve as targets of selection. The answer is "it depends." There are many different kinds of assemblages of individuals ("groups"), some of which do and others of which do not qualify as targets of selection. At one time I classified groups on the basis of size and geographic relationship (Mayr 1986), but this turned out not to be a productive approach. However, there is another approach that usually produces clear-cut results. It is obvious that a group, the selective value of which (when in isolation) is simply the arithmetic mean of the fitness values of the composing individuals, is not a target of selection. If such a group is particularly successful, it is due to the superior fitness of the composing individuals. This kind of group has often been included in theories of group selection. However, this false or "soft" group selection is not group selection at all. The fitness of such a group is the arithmetic mean of the fitness of the composing individuals. In contrast, if, owing to the interaction of the composing individuals or owing to a division of labor or other social actions, the fitness of a group is higher or lower than the arithmetic mean of the fitness values of the composing individuals, then the group as a whole can serve as an object of selection. I call this *hard group selection*. Interestingly, this was already appreciated by Darwin in a discussion of groups of primitive humans (Darwin 1871). Such hard group selection, a prerequisite for the explanation of human ethics, is still controversial (Sober and Wilson 1998).

It is sometimes difficult to decide whether the success of a particular group is due to soft or hard group selection. However, when a group of ground squirrels is particularly successful, because it has an efficient system of sentinels warning the group of approaching predators, it is clearly hard group selection. This is also the case when a pride of lionesses splits up to block the escape route of an intended victim. The success of surprise attacks by chimpanzees on members of neighboring troupes depends on the well-organized strategy of the attackers. In all such cases, the successful group acts as a unit and is, as a whole, the entity favored by selection. Such groups often consist of close relatives and such selection is actually kin selection. And kin selection is really individual selection.

No other potential object of selection has been as frequently the source of argument as the group. From the synthesis to the 1960s no evolutionist was a champion of group selection. It is not supported, indeed not even mentioned, by Dobzhansky (1937) or in my widely used text (Mayr 1963). It is not listed in the index of either volume. I fail to find a whole-hearted adoption of group selection in any other publication in contemporary evolutionary biology. Group selection is upheld only in some publications in behavioral biology and ecology. Konrad Lorenz frequently stated that some trait was favored by selection because it was "for the good of the species." The ecologists also tended to typological thinking, and one finds frequent references in the ecological literature (Allee, Emerson, Brereton, etc.) that amount to a support for group selection. These statements in general were ignored in the evolutionary literature. This all might well have changed when Wynne-Edwards published in 1962 a vigorous promotion of group selection. He claimed that in animals, particularly in birds, and specifically in red grouse, many life history traits had been acquired by group selection. This claim was at once, point by point, vigorously refuted by David Lack in a superb analysis (1966).

Lack was not the only one to reject the group selection thesis. G. C. Williams devoted an entire book (*Adaptation and Natural Selection*, 1966) to this purpose. He singled out for special attention (pp. 239–249) the claims made by Wynne-Edwards. Together with Lack's refutation this was the end of Wynne-Edwards's ill-founded claims. Williams's refutation of group selection was widely considered authoritative and accepted in the ensuing thirty-five years as the basis of most discussions of the group selection problem. Unfortunately, however, his presentation was seriously flawed in several important ways.

Williams apparently had some difficulty defining the term "group." Eventually he decided that a group, to qualify as an object of group selection, must have a *biotic adaptation*. Individuals have organic adaptations, but Williams coined the term biotic adaptation for those presumed adaptations of groups that qualify them to serve as objects of selection. He devotes chapters 5–8 (pp. 125–250) to test one possible biotic adaptation after the other to determine whether it qualifies under his definition. He finally concluded that none of the putative biotic adaptations qualifies and therefore group selection does not occur. But Williams's definition of biotic adaptation excludes numerous groups now considered legitimate objects of natural selection. The presence of biotic adaptations therefore was not a suitable classification.

What other criterion could possibly serve as a suitable criterion? Going back to Darwin (1871), I finally found that there was indeed one criterion by which to distinguish between groups that are potential objects of selection and those that are not (Mayr 1990). I called these two groups *casual groups* and *social groups*. Casual groups, as indicated by the name, are accidental associations of individuals, such as most flocks of starlings and schools of fishes. Their composition may change from hour to hour, and the mean fitness value of the casual group equals the arithmetic mean of the fitness value of the members of the group. If a herd of five deer consists of three slow ones and two fast ones, the fitness value of the herd would change drastically if predators killed the three slow ones. Casual groups as such are never the object of selection. The individuals of whom they are composed, however, are.

Social groups may have a fitness value that exceeds the arithmetic mean of the values of its members. The social cohesion of such a group results in all sorts of cooperation that increases its fitness in interaction with competing groups. Most social groups have a family as nucleus. To this may be added more distant relatives, such as grandchildren, cousins, nephews, nieces, uncles, aunts, etc. They all have known each other from birth on and were raised in a spirit of reciprocal helpfulness. This includes fighting together against outsiders, sharing the discovery of new sources of food and water, joint defense of caves and territories, and similar cooperative activities. Such a cohesive social group has a fitness value that considerably exceeds the arithmetic mean of the fitness values of its individual members. Darwin (1871), with his impeccable intuition, saw this clearly, and so did other authors after him. Williams (p. 116) quotes Ashley Montague as saying: "We begin to understand then that evolution is a process which favors cooperating rather than disoperating groups and that 'fitness' is a function of the group as a whole than [of] separate individuals." But Williams curiously refused to accept this as a case of group selection, because he attributed the success of such altruistic groups entirely to the characteristics of individuals (p. 117). Williams's failure to appreciate that social groups may have an entirely different fitness value from casual groups, resulted in a considerable confusion in the evolutionary literature.

Williams's long analysis failed to demonstrate that selection of social groups does not occur. On pp. 239–249, Williams refutes quite effectively the claims made by Wynne-Edwards (1962) in favor of group selection of casual groups, but he fails to demonstrate that group selection does not occur under different circumstances. Yet, this led him to a total rejection of group selection. This rejection was at first widely accepted, but during the last thirty-five years an increasing number of authors have acknowledged the potential of social groups for group selection. As Darwin pointed out, its existence is of great importance for the development of intragroup altruism. The final conclusion, widely accepted in evolutionary biology, is that casual groups are never an object of selection, but social groups, as cohesive units, may indeed be a target of selection. To qualify as a potential object of selection, a social group must be clearly delimited and compete with other such social groups.

Borrello (2003) recently attempted to restore the validity of Williams's arguments. However, this attempt was not successful because Borrello made the same mistakes as Williams. He did not realize that there are different kinds of groups, some of which may be legitimate objects of natural selection (social groups) while others are definitely not (casual groups).

Selection at higher levels

There has been much argument about whether there is, or is not, such a phenomenon as species selection. In the early post-Darwinian period when thinking about selection was rather confused, it was often said that such and such a character had evolved because it was "good for the species." This was quite misleading. The selected character had been favored because it benefited certain individuals of a species and had gradually spread to all others. The species as an entity does not answer to selection.

Selection that simultaneously affects different levels in the hierarchy of evolving entities has been referred to as multilevel selection. In most of these cases, one species is victorious in the struggle for existence between two competing species, but the actual selection takes place at the level of the individuals of which the two species populations are composed. As competitors for the same resource, they act as if they were members of a single species population and members of the inferior species preferably will be eliminated. Even in such cases of seeming species selection, individuals are the primary target of selection. If the object of selection, for example individual and species, belongs simultaneously to two different categorical levels, one speaks of "levels of selection."

There is, of course, no question that one species can cause the extinction of another species. The introduction of the Nile perch into Lake Victoria in Africa has resulted there in the extinction of several hundred endemic species of cichlid fishes. The parasitic cowbird almost exterminated the Kirtland's warbler in northern Michigan until drastic cowbird eradication procedures in the breeding range of Kirtland's warbler were adopted. Darwin described in 1859 the extermination of many native New Zealand species of animals and plants as a result of the introduction of competing species from England. The competitors were by no means always close relatives. In spite of all these examples, I hesitate to use the term species selection and prefer to call such events *species turnover* or *species replacement* because the actual selection takes place at the level of competing individuals of the two species. It is individual selection discriminating against the individuals of the losing species that causes the extinction. The result, however, is the survival of one of the two species and the extinction of the other.

Some authors have also suggested recognizing even higher levels of selection such as family selection or clade selection, but in no case are these entities as such the object of selection. Selection in these cases always takes place at the level of individuals. However, the stem mother of a new clade may supply genes to this clade that affect the fitness of all the individuals of the clade. Some authors like to refer to such cases as clade selection.

Terms for the objects of selection

A number of terms have been suggested for the entity favored by selection, but all of them, as I will show, are equivocal or saddled with the misleading meaning of their former everyday usage.

Unit of selection

This term was introduced in 1970 by Lewontin to designate the object of selection. In science as well as in daily life, the term unit usually means some measurable entity. We have units of length, weight, and time, and we have electrical units like volt, watt, ohm, etc. Clearly, unit of selection does not refer to this kind of unit. Occasionally, we also use the word unit for concrete entities – for example, "The president sent several units of marines to the area of the disturbances." The term unit of selection was adopted by many authors, but many others found it so unsuitable that they introduced new terms. Owing to its ambiguity, the term unit has been used less and less frequently in recent years, and "object of selection" is used instead.

Replicator

Dawkins, the author of the term, states, "We may define a replicator as any entity in the universe which interacts with its world, including other replicators in such a way that copies of itself are made" (Dawkins 1978). He also states that "a DNA molecule is the obvious replicator." In other words, replicator selection is essentially a new word for gene selection. One of the advantages of his term, says Dawkins, is that it automatically preadapts our language to deal "with non-DNA forms of evolution such as may be encountered on other planets." This strikes me as a rather curious excuse for introducing a new term into science.

With the phenotype of the individual rather than the gene being the target of selection, the term replicator becomes irrelevant. The term is, of course, in complete conflict with the basic Darwinian thought. What is important in selection is the abundant production of new phenotypes to permit the species to keep up with possible changes in the environment. This is made possible by meiosis and sexual reproduction. The replication of DNA has nothing to do with this. To be sure, Mendel's discovery of the constancy of genes, confirmed by all subsequent work in genetics and molecular biology, is a very efficient way to achieve rapid and unambiguous evolutionary change, and it refuted the inheritance of acquired characters. But such constancy is not necessary for selection, because Darwin's acceptance of an inheritance of acquired characters and a direct effect of the environment were compatible with natural selection. He did not demand complete constancy of the genetic material. Because the gene is not an object of selection (there are no naked genes), any emphasis on precise replication is irrelevant. Evolution is not a change in gene frequencies, as is claimed so often, but a change of phenotypes, in particular the maintenance (or improvement) of adaptedness and the origin of diversity. Changes in gene frequency are a result of such evolution, not its cause. The claim of gene selection is a typical case of reduction beyond the level where analysis is useful.

Vehicle

In due time Dawkins (1978) realized that the individual reproducing organism did have a role in the selection process. But being a gene selectionist, he saw this role only as the function to serve as a transport mechanism for genes. Therefore, he introduced the term *vehicle* for individuals. Doing so, he missed the decisive point that the phenotype is far more than a vehicle for the genotype. The term vehicle altogether fails to bring out the important role of the phenotype in the process of selection.

Interactor

Hull (1980) realized the unsuitability of the term vehicle because he appreciated that the object of selection acts "as a cohesive whole with its environment." To stress this interaction he proposed the term *interactor* "as an entity that directly interacts as a cohesive whole with its environment in such a way that replication [he meant reproduction] is differential." The term interactor has a number of weaknesses. One is the stress on constancy during replication while omitting any references to the production of variation during meiosis and reproduction. More serious is the fact that interactor is not a specific term for the object of selection. Every cell is an interactor; every organ of an organism interacts with the other organs, species interact, and so do classes of individuals such as the two sexes. Also, interacting is not conspicuous during the process of elimination that results in natural selection. In biology, interaction is far more pertinent to functional than to evolutionary biology. When one hears the word interactor, one's first thought would never be natural selection.

Target of selection

For many years I used the term target of selection for the object of selection. The more I realized, however, that natural selection is usually mostly an elimination process, the more I realized that the eliminated individuals were the real target of the selection process and that it was rather misleading to call those that remained the target of selection. What was needed was a more specific term.

Meme

Dawkins (1982) introduced the term "meme" for entities subject to selection in cultural evolution. It seems to me that this word is nothing but an unnecessary synonym of the term "concept." Dawkins apparently liked the word meme owing to its similarity to the word gene. In neither his definition nor the examples illustrating what memes are, does Dawkins mention anything that would distinguish memes from concepts. Concepts are not restricted to an individual or to a generation, and they may persist for long periods of time. They are able to evolve.

Selection of and selection for

Perhaps the two most important questions one can ask about selection are "selection of?" and "selection for?" as Sober (1984) perceptively pointed out. The question "selection of?" asks what the particular entity is that is selected – in other words, what entity has a superior survival probability to reproduce and to reproduce successfully? None of the terms discussed on the preceding pages seemed particularly suitable for this purpose. So the term *object of selection* would still seem to be the most appropriate designation, and it is, indeed, the one now most frequently used.

A trait that contributes to the fitness of an object of selection might be at almost any level of biological organization from the base pairs to the species. Very often it is a particular gene. But such a gene, being part of a genotype, is not an independent object of selection. This has often been confused in discussions of this subject.

The current status of natural selection

Darwin's theory of natural selection has been totally victorious after the complete refutation of typology and teleology. And yet, as I have shown, it is somewhat modified from Darwin's original theory. The implicit conflict between natural selection and random variation, which dominated the controversy between Dawkins and Gould, can now be seen as a cooperative process. No selection can take place without variation, and variation is meaningless without subsequent selection (elimination). The seeming antagonism between variation and selection can now be interpreted as a constructive process. Selection is not strictly a selection of the best, but largely an elimination of inferior members of the population. This explains the somewhat unexpected departure of new evolutionary developments, giving rise to evolutionary novelties.

Furthermore natural selection is now seen as two very different processes, natural selection proper (survival selection) and sexual selection (selection for reproductive success). In some populations or some periods in the life cycle, sexual selection may be more important than survival selection.

LITERATURE CITED

Borrello, M. E. 2003. Synthesis and selection: Wynne-Edwards' challenge to David Lack. *Journal of the History of Biology*, 36:531–566.

Brandon, R. N. 1990. *Adaptation and Environment*. Princeton: Princeton University Press.

Carson, H. L. 2002. Female choice in *Drosophila*: Evidence from Hawaii and implications for evolutionary biology. *Genetics*, 116:383–393.

Darwin, C. 1859. *On the Origin of Species by Means of Natural Selection or the Preservation of Favored Races in the Struggle for Life*. London: John Murray [1964, Facsimile of the First Edition; Cambridge, MA: Harvard University Press].

Darwin, C. 1871. *The Descent of Man*. London: Murray.

Dawkins, R. 1976. *The Selfish Gene*. Oxford: Oxford University Press.

Dawkins, R. 1978. Replicator selection and the extended phenotype. *Zeitschrift für Tierpsychologie*, 47:61–76.

Dawkins, R. 1982. *The Extended Phenotype: The Gene as the Unit of Selection*. Oxford: Freeman.

Dobzhansky, T. 1937. *Genetics and the Origin of Species*. New York: Columbia University Press.

Eberhard, W. 1996. *Female Control. Sexual Selection by Cryptic Female Choice*. Princeton: Princeton University Press.

Endler, J. A. 1986. *Natural Selection in the Wild*. Princeton: Princeton University Press.

Futuyma, D. J. 1999. *Evolutionary Biology*, 3rd ed. Sunderland, MA: Sinauer Associates.

Gould, S. J., 1989. *Wonderful Life: The Burgess Shale and the Nature of History*. New York: W. W. Norton.

Hamilton, W. D. 1964. The genetic evolution of social behavior. *Journal of Theoretical Biology*, 7:1–52.

Hull, D. 1980. Individuality and selection. *Annual Review of Ecology and Systematics*, 11:311–332.

Lack, D. 1966. *Population Studies of Birds*. Oxford: Clarendon Press.

Lerner, M. 1954. *Genetic Homeostasis*. Edinburgh: Oliver and Boyd.

Lewontin, R. 1970. The units of selection. *Annual Review of Ecology and Systematics*, 1:1–18.

Lloyd, E. 1992. Units of selection. In *Keywords in Evolutionary Biology*, E. F. Keller and E. A. Lloyd (eds.). Cambridge, MA: Harvard University Press, pp. 334–340.

Mather, K. 1943. Polygenic inheritance and natural selection. *Biological Reviews*, 18:32–64.

Mayr, E. 1959. Where are we? *Cold Spring Harbor Symposia on Quantitative Biology*, 24:1–14.

Mayr, E. 1963. *Animal Species and Evolution*. Cambridge, MA: Harvard University Press.

Mayr, E. 1976. *Evolution and the Diversity of Life*. Cambridge, MA: Harvard University Press.

Mayr, E. 1977. The study of evolution historically viewed. In *The Changing Scenes in the Natural Sciences 1776–1976*, C. F. Goulden (ed.). Special Publications 12:39–58. Philadelphia: Academy of Natural Sciences.

Mayr, E. 1986. The philosopher and the biologist. Review of *The Nature of Selection: Evolutionary Theory in Philosophical Focus* by Elliott Sober, 1984. *Paleobiology*, 12:235–239.

Mayr, E. 1990. Myxoma and group selection. *Biologisches Zentralblatt*, 109:453–457.

Mayr, E. 1997. The objects of selection. *Proceedings of the National Academy of Sciences*, 94:2091–2094.

Mayr, E. 2001. *What Evolution Is*. New York: Basic Books.

Sober, E., and Lewontin, R. 1982. Artifact, cause, and genic selection. *Philosophy of Science*, 49:157–180.

Sober, E. 1984. *The Nature of Selection: Evolutionary Theory in Philosophical Focus*. Cambridge, MA: MIT Press.

Sober, E., and Wilson, D. S. 1998. *Unto Others*. Cambridge, MA: Harvard University Press.

Wallace, B. 1954. Coadaptation and the gene arrangements of *Drosophila pseudoobscura*. *IUBS Symposium on Genetics of Population Structures*, pp. 67–100.

Wallace, B., King, J. C., Madden, C. V., Kaufmann, B., and McGunnigle, E. C. 1953. An analysis of variability arising through recombination. *Genetics*, 38:272–307.

West-Eberhard, M. J. 2003. *Developmental Plasticity and Evolution*. New York: Oxford University Press.

Williams, G. C. 1966. *Adaptation and Natural Selection*. Princeton: Princeton University Press.

Williams, G. C. 1996. *Adaptation and Natural Selection* [with new preface]. Princeton: Princeton University Press.

Wilson, E. O. 1975. *Sociobiology: The New Synthesis*. Cambridge, MA: Harvard University Press.

Wimsatt, W. C. 1980. Reductionist research strategies and their biases in the units of selection controversy. In *Scientific Discovery*, T. Nickles (ed.). Dordrecht: Reidel.

Woltereck, R. 1909. Weitere experimentelle Untersuchungen über Artveränderung, speziell über das Wesen quantitativer Artunterschiede bei Daphnien. *Verhandlungen der Deutschen Zoologischen Gesellschaft*, 19:10–173.

Wynne-Edwards, V. C. 1962. *Animal Dispersion in Relation to Social Behaviour*. Edinburgh and London: Oliver and Boyd.

9

Do Thomas Kuhn's Scientific Revolutions Take Place?[1]

ACCORDING TO THOMAS KUHN'S CLASSIC THESIS (1962), science advances through occasional scientific revolutions, separated by long periods of "normal science." During a scientific revolution, a discipline adopts an entirely new "paradigm," which dominates the ensuing period of normal science. The key concept in Kuhn's discussion of scientific revolutions is the occurrence of such paradigm shifts. One of Kuhn's critics has claimed that Kuhn had used the term paradigm in at least twenty different ways in the first edition of his book. For the most important one, Kuhn later introduced the term "disciplinary matrix." A disciplinary matrix (paradigm) is more than a new theory; it is, according to Kuhn, a system of beliefs, values, and symbolic generalizations. There

---

[1]   Previously published [Mayr (1994)].

is a considerable similarity between Kuhn's disciplinary matrix and terms of other philosophers such as "research tradition."

Revolutions (paradigm shifts) and periods of normal science are only some aspects of Kuhn's theory. One other one is a supposed incommensurability between the old and the new paradigm. Hoyningen-Huene (1993) has presented an excellent analysis of Kuhn's views, including various changes after 1962.

Few publications in the history of the philosophy of science have created as great a stir as Kuhn's *The Structure of Scientific Revolutions*. Many authors were able to confirm his conclusions; perhaps more others were unable to do so. There are numerous more or less independent aspects of Kuhn's thesis, but they cannot be discussed profitably without looking at concrete cases. It is necessary to study particular sciences at particular periods and ask whether theory change did or did not follow Kuhn's generalizations. I have therefore analyzed a number of major theory changes in biology.

For instance, in macrotaxonomy, the science of animal and plant classification, we can distinguish an early period from the herbalists (sixteenth century) to Carl Linnaeus, when most classifications were constructed by logical division. The nature of the changes, made from one classification to another, depended on the number of classified species and on the weighting of different kinds of characters. This type of methodology is referred to as *downward classification*. In due time, it was realized that this was really a method of identification and it was supplemented by a very different method – *upward classification* – consisting of the arrangement of ever-larger groups of related species in a hierarchical fashion. (The method of downward classification continued to exist side by side, being used in keys in all taxonomic revisions and monographs and in field identification guides.) Upward classification was first used by some herbalists and later by Pierre Magnol (1689) and by Michel Adanson (1772). This method did not begin to be generally adopted until the last quarter of the eighteenth century. There was no revolutionary replacement of one paradigm by another one (Mayr 1982: chapter 5) because both continued to exist, although with different objectives.

One would have expected that the adoption of Charles Darwin's theory of common descent in 1859 would have produced a major taxonomic revolution, but this was not the case for the following reason. In upward classification, groups are recognized on the basis of the greatest number of shared characters. Not surprisingly, the taxa thus delimited consisted usually of descendants of the nearest common ancestor. Hence, Darwin's theory supplied the justification for the method of upward classification but the theory of common descent did not result in a scientific revolution in taxonomy.

Let us now look at another field, evolutionary biology. The simple picture of the biblical story began to lose credence by the end of the seventeenth century. In the eighteenth century, when the long duration of geological and astronomical time was beginning to be appreciated, when the biogeographic differences of the different parts of the world were discovered, when an abundance of fossils were described, etc. (Mayr 1982), various new scenarios were proposed, including repeated creations, all of them, however, operating with new origins. All of them existed side by side with the biblical story of creation, which was still supported by the vast majority. The first to seriously undermine these views was Buffon (1749), many of whose ideas were in complete opposition to the essentialistic-creationist world picture of his time (Roger 1997). Indeed it was from his ideas that the evolutionary thinking of Denis Diderot, J. F. Blumenbach, J. G. Herder, Jean-Baptist Lamarck, and others was derived. When in 1800 Lamarck proposed the first theory of genuine gradual evolution, he made few converts; he did not start a scientific revolution. Furthermore, those who followed him as evolutionists, like Étienne Geoffroy and Robert Chambers, differed widely in many respects from Lamarck and from each other. He certainly had not effected the replacement of one paradigm by a new one.

No one can deny that Darwin's *Origin of Species* (1859) produced a genuine scientific revolution. Indeed it is often called the most important of all scientific revolutions. Yet, it does not at all conform to Kuhn's specifications of a scientific revolution. The analysis of the Darwinian revolution encounters considerable difficulties because Darwin's paradigm actually consisted of a whole package of theories, five of which are most important (Mayr 1991: chapter 6). Matters become much clearer if one speaks of Darwin's first and second scientific revolutions. The first one consisted of the acceptance of evolution by common descent. This theory was revolutionary in two respects. First, it replaced the concept of special creation, a supernatural explanation, by that of gradual evolution, a natural, material explanation. And second, it replaced the model of straight-line evolution, adopted by earlier evolutionists, by that of branching descent, requiring only a single episode of origin of life. This was finally a persuasive solution for what numerous authors, from Linnaeus on (and earlier) had attempted, to find a "natural" system. It rejected all supernatural explanations. It furthermore involved depriving humans of their unique position and placing them in the animal series. Common descent was remarkably rapidly adopted and formed perhaps the most successful research program of the immediate post-Darwinian period. The reason is that it fitted so well into the research interests of morphology and systematics, supplying a theoretical explanation of previously discovered empirical evidence, such as the Linnaean hierarchy and the archetypes of Richard Owen and Karl Ernst von Baer. It did not involve any drastic shift of a paradigm. Furthermore, if one were to accept the period from Georges Louis Buffon (1749) to the *Origin* (1859) as a period of normal science, one would have to deprive a number of minor revolutions, which took place within this period, of their revolutionary status. This includes the discovery of the great age of the earth, of extinction, of the replacement of the *scala naturae* by morphological types, of biogeographic regions, of the concreteness of species, etc. All of these were necessary prerequisites for Darwin's theories and could be included as components of the first Darwinian revolution, shifting the beginning of the Darwinian revolution back to 1749 (Mayr 1972).

The second Darwinian revolution (Mayr 1991) was caused by the theory of natural selection. Although proposed and fully explained in 1859, it encountered such solid opposition owing to its conflict with five prevailing ideologies that it was not generally accepted until the evolutionary synthesis of the 1930s–1940s. And in France, Germany, and some other countries there is still considerable resistance to it even at the present time. When did this second Darwinian revolution take place? – when it was proposed (1859) or when it was broadly adopted (1940s)? Can one consider the period from 1859 to the 1940s a period of normal science? Actually a considerable number of minor scientific revolutions took place in this period, such as the refutation of an inheritance of acquired characters (Weismann 1883), the rejection of blending inheritance (Mendel 1866), the development of the biological species concept (E. B. Poulton, K. Jordan, E. Mayr, etc.), the discovery of the source of genetic variation (mutation, genetic recombination, diploidy), the appreciation of the importance of stochastic processes in evolution (J. T. Gulick, Sewall Wright), the founder principle (E. Mayr), the proposal of numerous genetic processes of evolutionary consequence, etc. Many of these had indeed a rather revolutionary impact on the thinking of evolutionists but without any of the Kuhnian attributes of a scientific revolution.

After the general adoption of the synthetic theory, let us say from 1950 on, modifications of almost all aspects of the paradigm of the synthesis were proposed and some were adopted. Nevertheless, there can be little doubt that throughout the period from 1800 to the present there were periods of relative quiet in evolutionary biology and other periods of rather vigorous change and controversy. In other words, neither the Kuhnian image of well-defined short revolutions and intervening long periods of normal science is correct, nor is that of his most extreme opponents, of slow, steady, even progress.

Perhaps the most revolutionary development of biology in the twentieth century was the rise of molecular biology. It resulted in a new field, with new scientists, new problems, new experimental methods, new journals, new textbooks, and new culture heroes, but, as John Maynard Smith has stated correctly, conceptually the new field was nothing but a smooth continuation of the developments in genetics preceding 1953. There was no revolution during which the previous science was rejected. There were no incommensurable paradigms. Rather it was the replacement of coarse-grained by fine-grained analysis and the development of entirely new methods. The rise of molecular biology was revolutionary, but it was not a Kuhnian revolution.

It would be interesting, but has not yet been done, to look at breakthroughs in various other fields of biology and see to what extent they qualify as revolutions, and whether they led to the replacement of one paradigm by another, and how much time it took before the replacement was completed. For instance, was the origin of ethology (Konrad Lorenz, Niko Tinbergen) a scientific revolution? In what respects was the proposal of the cell theory (Th. Schwann, M. J. Schleiden) a scientific revolution?

The same new theory may be far more revolutionary in some sciences than in others. Plate tectonics supplies a good illustration. That this theory had a revolutionary, one might almost say cataclysmic, effect on geology is obvious. But what about biology? As far as avian distributions are concerned, the historical narrative inferred before plate tectonics (Mayr 1946) had to be changed hardly at all (a North Atlantic connection in the early Tertiary is the only exception) as a result of the adoption of plate tectonics. To be sure, avian distribution in Australonesia did not agree at all with plate tectonic reconstructions, but later geological work showed that the geological reconstructions were faulty, while the revised construction fitted the biological postulates quite well. That there must have been a Pangaea in the Permian-Triassic had been postulated by paleontologists long before the proposal of plate tectonics. In other words, the interpretation of the history of life on earth was not nearly as much affected by the acceptance of plate tectonics as was that of geology.

Virtually every author who has attempted to apply Kuhn's thesis to theory change in biology has found that it is not applicable in his field. This conclusion is inevitable when one looks at the so-called revolutions in biology described in the above given case histories. Even in the cases in which there was a rather revolutionary change, it did not at all take place in the form described by Kuhn. There are a number of pronounced differences. First of all, there is no clear-cut difference between revolutions and "normal science." What one finds is a complete gradation between minor and major theory changes. A number of minor "revolutions" take place even in any of the periods that Kuhn might designate as "normal science." Up to a point this is also admitted by Kuhn (Hoyningen-Huene 1993), but did not induce him to abandon his distinction between revolutions and normal science.

The introduction of a new paradigm by no means always results in the immediate replacement of the old one. As a result, the new revolutionary theory may exist side by side with the old one. In fact, as many as three or four paradigms may coexist. For instance, after Darwin had proposed natural selection as the mechanism of evolution, saltationism, orthogenesis, and Lamarckism competed with selectionism for the next eighty years (Bowler 1983). It was not until the evolutionary synthesis of the 1940s that these competing paradigms lost their credibility.

Kuhn makes no distinction between theory changes caused by new discoveries and such that are the result of the development of entirely new concepts. Changes caused by new discoveries usually have much less of an impact on the paradigm than conceptual upheavals. For instance, the ushering in of molecular biology through the discovery of the structure of the double helix had only minor conceptual consequences. There was virtually no paradigmatic change during the transition from genetics to molecular biology, as has been pointed out by Maynard Smith and others.

The major impact of the introduction of a new paradigm may be a massive acceleration of research in the area. This is particularly well illustrated by the explosion of phylogenetic research after the proposal by Darwin of the theory of common descent. In comparative anatomy as well as in paleontology, much of the research after 1860 was directed to the search for the phylogenetic position of specific taxa, particularly primitive and aberrant ones. There are many other instances in which remarkable discoveries had relatively little impact on the theory structure of the field. The unexpected discovery by Meyen and Robert Remak that new cells originate by the division of old cells and not by the conversion of the nucleus into a new cell had remarkably little impact. As far as genetic theory is concerned, likewise, the discovery that the genetic material is nucleic acids rather than proteins did not lead to a paradigm shift.

The situation is somewhat different with the development of new concepts. When Darwin's theorizing forced including humans in the tree of common descent, it indeed caused an ideological revolution. On the other hand, as was correctly emphasized by Popper, Mendel's new paradigm of inheritance did not. Changes in concepts have far more impact than new discoveries. For instance, the replacement of essentialistic by population thinking had a revolutionary impact in the fields of systematics, evolutionary biology, and even outside of science (in politics). This shift had a profound effect on the interpretation of gradualism, speciation, macroevolution, natural selection, and racism. The rejection of cosmic teleology and of the authority of the Bible have had equally drastic effects on the interpretation of evolution and adaptation.

The impact of a revolutionary new concept or discovery on the prevailing paradigm is highly variable. In the case of Darwin's theory of natural selection, the ideological commitment of the preceding paradigm to essentialism, theism, teleology, and physicalism necessitated not only the most profound revolution ever produced by a new theory but also the longest period of delay (Mayr 1991).

The publication in 1859 of Darwin's *Origin of Species* was unique in representing a multiple scientific revolution. I am referring to the very special case of the simultaneous proposal of several revolutionary theories, such as that of common descent and of natural selection. These are really two independent scientific revolutions and either one can exist without the other. The enthusiastic acceptance of the theory of common descent and the virtual nonacceptance of the theory of natural selection in the first eighty years after 1859 definitely proves this independence. The reason for the difference in reception is that common descent was rather easily accommodated in the thinking of the period while natural selection was not.

Finding virtually no confirmation of Kuhn's thesis in a study of theory changes in biology inevitably forces us to ask what induced Kuhn to propose his thesis? Because much explanation in physics deals with the effects of universal laws, such as we do not have in biology, it is indeed possible that explanations involving universal laws are subject to Kuhnian revolutions. But we must also remember that Kuhn was a physicist and that his thesis, at least as presented in his early writings, reflects the essentialistic-saltationistic thinking so widespread among physicists. Each paradigm was at that time for Kuhn of the nature of a Platonic eidos or essence and could change only through its replacement by a new eidos. Gradual evolution would be unthinkable in this conceptual framework. The variations of an eidos are only "accidents," as it was called by the scholastic philosophers, and therefore the variation in the period between paradigm shifts is essentially irrelevant, merely representing "normal science." The picture of theory change that Kuhn painted in 1962 was congenial to the essentialistic thinking of physicalists. However, it is incompatible with the gradualistic thinking of a Darwinian. Therefore, it is not surprising that the Darwinian epistemologists introduced an entirely different conceptualization for theory change in biology, usually referred to as evolutionary epistemology.

The principal thesis of Darwinian evolutionary epistemology is that science, as reflected in its currently accepted epistemology, advances very much as does the organic world during the Darwinian process. Epistemological progress thus is characterized by variation and selection.

One can perhaps draw the following conclusions from these observations:

(1) There are indeed major and minor revolutions in the history of biology.

(2) Yet even the major revolutions do not necessarily represent sudden, drastic paradigm shifts. An earlier and the subsequent paradigm may coexist for long periods. They are not necessarily incommensurable.

(3) Active branches of biology seem to experience no periods of "normal science." There is always a series of minor revolutions between the major revolutions. Periods without such revolutions are found only in inactive branches of biology, but it would seem inappropriate to call such inactive periods "normal science."

(4) The descriptions of Darwinian evolutionary epistemology seem to fit theory change in biology better than Kuhn's description of scientific revolutions. Active areas of biology experience a steady proposal of new conjectures (Darwinian variation) and some of them are more successful than others. One can say that these are "selected" until replaced by still better ones.

(5) A prevailing paradigm is likely to be more strongly affected by a new concept than by a new discovery.

LITERATURE CITED

Bowler, P. J. 1983. *The Eclipse of Darwinism*. Baltimore: John Hopkins Press.
Darwin, C. 1859. *On the Origin of Species by Means of Natural Selection or the Preservation of Favored Races in the Struggle for Life*. London: John Murray.

[1964. Facsimile of the First Edition, Cambridge, MA: Harvard University Press.]

Hahlweg, K., and C. A. Hooker. 1989. *Issues in Evolutionary Epistemology*. Albany: State University of New York Press.

Hoyningen-Huene, P. 1993. *Reconstructing Scientific Revolutions: Thomas S. Kuhn; Philosophy of Science*. Chicago: Chicago University Press.

Kuhn, T. 1962. *The Structure of Scientific Revolutions*. Chicago: Chicago University Press.

Mayr, E. 1946. History of North American bird fauna. *Wilson Bulletin*, 58:3–41.

Mayr, E. 1972. The nature of the Darwinian revolution. *Science*, 176:981–989.

Mayr, E. 1982. *The Growth of Biological Thought*. Cambridge, MA: Harvard University Press.

Mayr, E. 1991. *One Long Argument*. Cambridge, MA: Harvard University Press.

Mayr, E. 1994. The advance of science and scientific revolutions. *Journal of the History of Behavioral Sciences*, 30:328–334.

Roger, J. 1997 [1989]. *Buffon. A Life in Natural History*. Ithaca: Cornell University Press.

Weismann, A. 1883. *Über die Vererbung*. Jena: Gustav Fischer.

IO

Another Look at the
Species Problem

T HE SPECIES, together with the gene, the cell, the individual, and the
local population, are the most important units in biology. Most research
in evolutionary biology, ecology, behavioral biology, and almost any other
branch of biology deals with species. How can one reach meaningful
conclusions in this research if one does not know what a species is and,
worse, when different authors talk about different phenomena but use
for them the same word – species? But this, it seems, is happening
all the time, and this is what is referred to as the species problem.
There is perhaps no other problem in biology on which there is as much
dissension as the species problem. Every year several papers, and even
entire volumes, are published, attempting to deal with this problem.

The species is indeed a fascinating challenge. In spite of the maturation of Darwinism, we are still far from having reached unanimity
on the origin of new species, on their biological meaning, and on the

delimitation of species taxa. The extent of the remaining confusion is glaringly illuminated by a recent book on the phylogenetic species concept (Wheeler and Meier 2000). From the discussions of some of the participating authors, it is quite evident that they are unaware of much of the recent literature. The result is great confusion. This induced me, contrary to earlier intentions, to write here once more about the species problem, even though I discussed the subject only quite recently (Mayr 1987, 1988, 1996, 2000). Unfortunately several authors of recent papers on the species problem had only a rather limited practical experience with species. They had never dealt with concrete taxonomic situations involving the rank (species or not?) of natural populations; in other words, they had no practical experience with actual species in nature. Their theorizing fails to provide answers for the practicing taxonomist. I am presumably well qualified to deal with this subject, having discussed the species problem in sixty-four books and scientific papers, published from 1927 to 2000. I also had to make decisions on species status when describing 26 new species and 473 new subspecies of birds. Furthermore, I had to make decisions on the rank of species level taxa in twenty-five generic revisions and faunistic reviews. Hence, there should be no doubt about my qualifications as a practicing systematist.

The reading of some recent papers on species has been a rather troubling experience for me. There is only one term that fits some of these authors: *armchair taxonomists*. Because they had never personally analyzed any species populations or studied species in nature, they lacked any feeling for what species actually are. Darwin already knew this when, in September 1845, he wrote to Joseph Hooker, "How painfully true is your remark that no one has hardly the right to examine the question of species who has not minutely described many" (Darwin 1887: 253). These armchair taxonomists tend to make the same mistakes that have been pointed out repeatedly in the recent literature. Admittedly, the relevant literature is quite scattered, and some of it may be rather inaccessible to a nontaxonomist. Yet, because the species concept is an important concept in the philosophy of science, every effort should be made to clarify it. I here attempt to present, from the perspective of a practicing systematist, a concise overview of the most important aspects of the "problem of the species."

The species is the principal unit of evolution. A sound understanding of the biological nature of species is fundamental to writing about evolution and indeed about almost any aspect of the philosophy of biology. A study of the history of the species problem helps to dispel some of the misconceptions (Mayr 1957, Grant 1994).

What is the nature of the problem?

There are a number of possible answers to this question. Is it perhaps that different kinds of organisms actually do have different kinds of species? It turns out that this is definitely the case because what is designated as species in asexually reproducing organisms (agamospecies) is indeed something quite different from the species of sexually reproducing organisms (see below). But one can also question whether even all species in sexually reproducing organisms are of only one kind.

Are there different kinds of species taxa?

No good comparative analysis is so far available in which species taxa were compared that differ drastically in their population structure. As an ornithologist, I am most familiar with bird species. They tend to become adapted to local conditions through producing geographic races (subspecies) (Mayr and Diamond 2001). The question can be raised whether this kind of geographically varying species is the same as the species of strictly host-specific, herbivorous insects where speciation is effected by the colonization of a new host species (usually by a more or less sympatric process of speciation). While a good percentage of bird species shows geographic variation and forms subspecies, thus becoming polytypic species, most host-specific herbivores seem to remain monotypic.

Speciation by colonization of a new host is always by budding, with the parental species remaining unchanged. It is quite unknown how many taxa of animals, plants, fungi, and protists have modes of speciation that might produce different kinds of species taxa. For instance, are polyploids a different kind of species from diploids? There is still a great deal of research to be done.

A major source of the species problem is that the word species has been used for two entirely different entities: for species concepts and for species taxa.

A *species concept*, as the word says, is the concept naturalists or systematists have of the role a species plays in nature. What kind of phenomenon do they have in mind, when they use the word species?

A *species taxon* is a population of organisms qualifying for recognition as a species taxon according to a particular species concept.

Species concept and species taxon are two entirely different phenomena, as is evident from these definitions. It creates a species problem when an author confounds the two phenomena. Therefore, let us analyze the several uses of the word species in more detail.

Species concepts

The typological species concept

Authors from Plato and Aristotle until Linnaeus and the early nineteenth century recognized species, *eide* (Plato) or kinds (Mills), on the basis of their difference. The word species conveyed the idea of a class of objects, the members of which share certain defining properties. Its description distinguished a given species from all other species. Such a class is constant, it does not change in time, and all deviations from the description of the class are merely accidents – that is, imperfect manifestations of its essence (eidos). Mill, in 1843, introduced the word "kind" for species, and philosophers have since occasionally used the term *natural kind* for species (as defined above), particularly after B. Russell and Quine had adopted it.

The current use of the term species for inanimate objects such as nuclear species or species of minerals reflects this classical typological concept. Up to the nineteenth century this was also the most practical species concept in biology. The naturalists were busy making an inventory of species in nature and the method they used for the discrimination of species was the identification procedure of downward classification (Mayr 1982, 1992a, 1992b). Species were recognized by their differences; they were kinds, and they were types. There were a number of different designations for species under this concept: Linnaean species, typological species, and morphological species. A limited amount of variation was acceptable under this typological concept and in recent years an increasing number of animal species were recognized by nonmorphological characters, such as behavior and pheromones. Most sibling species, for instance, did not fit the terminology of the morphological species concept. For this reason, perhaps, one should rather use the term phenotypic species concept, recognizable by the degree of their phenotypic difference.

At the time of Linnaeus, this concept was supported by three kinds of observations or considerations. First, by the principle of logic that variation has to be segregated into species. Second, by the observation of naturalists that organic variation consisted of species. There was no argument about what the species of birds were that one found in one's garden. And third, the Christian dogma that the variety of living nature consisted of the descendants of the pair of each kind created by God at the beginning. And so Linnaeus and his contemporaries had little difficulty in sorting organic individuals into species. Indeed, they applied this principle not only to living nature but also to inanimate entities such as minerals. A typological species is an entity that differs from other species by constant diagnostic differences, but it is subjective what one may consider a diagnostic difference. The so-called typological species concept is simply a biologically arbitrary means for delimiting species taxa. The results of this procedure are classes (natural kinds) not necessarily with the properties of biological species.

The criterion of species status in the case of the morphological (typological) species concept is the degree of phenotypic difference. According to this concept, a species is recognizable by an intrinsic difference reflected in its morphology, and it is this that makes one species clearly different from any and all other species. A species under this concept is a class recognizable by its defining characters. A museum or herbarium taxonomist who has to sort numerous collections in space and time and assign them to concrete and preferably clearly delimited taxa may find it most convenient to recognize strictly phenetic species in these cataloguing activities. I will presently point out the difficulties caused by this approach.

In the course of time, weaknesses in the typological concept developed. More and more often species were found in nature containing numerous conspicuously different intraspecific phena – that is, differences caused by sex, age, season, or ordinary genetic variation – that were so different from each other that members of the same population sometimes differed more strikingly from each other than generally recognized good species.

Conversely, in many groups of animals and plants extremely similar and virtually indistinguishable cryptic species were discovered that, where coexisting in nature, did not interbreed with each other but maintained the integrity of their respective gene pools. Such cryptic or *sibling species* certainly invalidate a species concept based on degree of difference. They occur at lesser or greater frequency in almost all groups of organisms (Mayr 1948), but are apparently particularly common among protozoans. Sonneborn (1975) eventually recognized fourteen sibling species under what he had originally considered a single species,

Paramecium aurelia. Many sibling species are genetically as different from each other as morphologically distinct species.

Perhaps the greatest weakness of the typological species concept is that it fails to answer the Darwinian "why?" question. It sheds no light on the reasons for the existence of discrete reproductively isolated species in nature. It tells us nothing about the biological significance of species. So-called morphological species definitions are nothing but man-made operational instructions for the demarcation of species taxa.

The biological species concept

An observation made by naturalists suggested an entirely different criterion on which to base species recognition. It was the realization that the individuals of a species form a reproductive community. Members of different species, even when they coexist at the same locality, do not ordinarily interbreed with each other. They are separated by an invisible barrier into reproductive communities. Each reproductively isolated community is called a *biological species*. The concept that bases species recognition on reproduction is called the *biological species concept* (BSC).

I define biological species as "groups of interbreeding natural populations that are reproductively (genetically) isolated from other such groups." The emphasis of this definition is no longer on the degree of morphological difference but rather on genetic relationship. In earlier definitions, I included the potential interbreeding of conspecific populations that were geographically separated. I now consider the word potential superfluous because "interbreeding" implies the possession of isolating mechanisms that permit the interbreeding of populations that are prevented from doing so by extrinsic barriers. The concept interbreeding thus includes the propensity for interbreeding. The word interbreeding indicates a propensity: a spatially or chronologically isolated population, of course, is not interbreeding with other populations, but it may have the propensity to do so when the extrinsic isolation is terminated. Species status is the property of populations, not of individuals. A population does not lose its species status when occasionally an individual belonging to it makes a mistake and hybridizes with another species.

It is very important to understand what the word concept means when combined with the word species. It conveys the meaning of the species in nature. A population or group of populations is a species according to the BSC if it forms a reproductive community and does not reproduce with members of other such communities. The BSC, thus defined, plays a concrete role in nature and differs in this respect from all those other so-called species concepts that are nothing but instructions, based on human judgment, on how to delimit species taxa. Every proposed so-called new species concept must be tested to see whether it really embodies a new meaning of the species in nature or is simply a new set of instructions for the delimitation of species taxa on the basis of a particular species concept.

This new interpretation of species of organisms emphasized that biological species are something very different from the natural kinds of inanimate nature. This was not fully understood until Darwin made it legitimate to ask "why?" questions in biology. To achieve a real understanding of the meaning of species, it was necessary to ask, Why are there species? Why do we not find in nature simply an unbroken continuum of similar or more widely diverging individuals? (Mayr 1988b). The reason, of course, is that each biological species is an assemblage of well-balanced harmonious genotypes. An indiscriminate interbreeding of all individuals in nature would lead to an immediate breakdown of these harmonious genotypes. The study of hybrids, with their reduced viability (at least in the F_2) and fertility, has demonstrated this abundantly. Consequently, there is a high selective premium on the acquisition of devices, now called *isolating mechanisms*, that would favor breeding with conspecific individuals and that would inhibit mating with nonconspecific individuals. This conclusion provides the true meaning of species. The species enables the protection of harmonious, well-integrated genotypes. It is this insight on which the BSC is based.

The BSC is most meaningful in local situations where different populations in reproductive condition are in contact with each other. The decision on which of these populations are considered species is not made on the basis of their degree of difference. They are assigned species status on a purely empirical basis – that is, on the observed criterion of presence or absence of interbreeding. Observations in the local situation have clearly demonstrated the superior reliability of the interbreeding criterion over that of degree of difference. This conclusion is supported by numerous detailed analyses of local biota. I refer, for instance, to the plants of Concord Township (Mayr 1992a), the birds of North America (Mayr and Short 1970), and the birds of Northern Melanesia (Mayr and Diamond 2001). In particular, there is no difficulty when there is a continuity of populations and gene flow results in genotypic cohesion of the assemblage of populations. It is this combination of interbreeding and gene flow that gives a biological species taxon its internal cohesion.

The BSC has a long history. It began with Buffon in 1749 (Sloan 1987) and continued with K. Jordan, E. Poulton, E. Stresemann, and B. Rensch. It is quite misleading to claim, as was done by some geneticists, that Dobzhansky was the author of BSC. A number of recent historians have credited me with the authorship of the BSC. This also is not correct. My merit was to propose a simple, concise definition that is now almost universally used in papers dealing with the BSC. But this definition, more than anything else, favored the acceptance of the BSC.

CRITICISM OF THE BSC. Why is the BSC, even though so widely adopted, still so often attacked? An analysis of numerous papers critical of the BSC leads me to the conclusion that the criticism is almost invariably due to a failure of the critics to make a clear distinction between the species category (species concept) and the species taxon. The BSC (and species definition) deals with the definition of the species category and with the concept on which it is based. This concept, protection of a harmonious gene pool, is strictly biological and of course has meaning only where a gene pool comes into contact with the gene pools of other species – that is, at a given locality at a given time (the nondimensional situation). Only where two natural populations meet in space and time can it be determined what is responsible for the maintenance of their integrity. There is never any doubt in sexually reproducing species what is the reproductive barrier. Two closely related sympatric species retain their distinction not because they are different in certain taxonomic characters, but because they are reproductively incompatible. The definition of monozygotic twins, as Simpson (1961) pointed out so rightly, provides a homologous causal equivalent. Two similar brothers are not monozygotic twins because they are so similar, but they are so similar because they are monozygotic twins. It is the concept of reproductive isolation that provides the yardstick for delimitation of species taxa, and this can be studied directly only in the nondimensional situation. However, because species taxa have an extension in space and time, the species status of noncontiguous populations must be determined by inference (see below).

Because I recently presented a detailed analysis of a number of criticisms of the BSC, I will not repeat myself but simply refer to that analysis (Mayr 1992a: 222–231). Here I answer only a few criticisms that have been made more recently.

Learning that the BSC reflects the nondimensional situation, Kimbel and Rak (1993: 466) concluded that it is a "failure of the Biological Species Concept to explain the temporal persistence criterion of individuality." This objection confuses the species concept with the delimitation of species taxa. One arrives at the definition of the species concept under the condition of nondimensionality, but species taxa have, of course, an extension in space and time: they are not newly created in every generation. The BSC presents us with the great advantage of providing a yardstick that permits us to infer which populations in space and time should be combined into one reproductively cohesive assemblage of populations and which others should be left out. As we shall presently see, none of the competing species concepts has such a criterion.

I want to emphasize particularly that *evolving* is not a species criterion, as has been claimed by a number of recent authors. Species do not differ in this respect from other living entities. Of course, every species is a product of evolution, but so is every population, every isolate, every species group, and every monophyletic higher taxon.

A population or a group of populations is a species under the BSC because it is a reproductive community and does not reproduce with other such species. The biological species, thus defined, plays a concrete role in nature and this species concept differs in this respect from all those other so-called species concepts that are nothing but instructions based on human judgment on how to delimit species taxa. Every newly proposed so-called new species concept must be tested to see whether it really embodies a new meaning of the species in nature or is simply a new set of instructions for the delimitation of species taxa.

DIFFICULTIES IN THE APPLICATION OF THE BSC. Species evolve, as does everything in living nature. Subspecies in the course of time may become incipient species and eventually full species. In every group of organisms, there are situations in which populations are at that interme-diate stage between "not yet species" and "already full species." As far as the biological species in birds is concerned this is true particularly for ge-ographically isolated populations (Mayr and Diamond 2001). The status of such populations can be determined only by inference. One must ask, does the phenotype of such populations indicate that they have reached species level? By necessity, the answer to this question will be subjective. But fortunately the percentage of arguable cases is small. Evolution is responsible for the fact that such borderline cases are encountered in the application of any species concept. For a more detailed discussion see Mayr (1988a, 1992, 1996).

Before going on to an analysis of more difficult situations, let me repeat that the BSC is inapplicable to asexual organisms, which form clones, not populations. Because asexual organisms maintain their genotype from generation to generation by not interbreeding with other organisms, they are not in need of any devices (isolating mechanisms) to protect the integrity and harmony of their genotype. In this I entirely agree with Ghiselin (1974).

Most of the criticisms of the BSC are directed against decisions made in applying the BSC in the delimitation of species taxa. Using the BSC as a yardstick in ordering contiguous interbreeding populations causes no difficulties. However, it would seem that the criterion of interbreeding cannot be applied in the delimitation of species wherever isolated populations are involved – populations isolated in either time or space. I have presented in great detail the reasoning used by the defenders of the BSC when assigning such populations to biological species (most recently in Mayr 1988a, 1988b, and 1992a). I will now summarize my arguments but refer to the cited publications for further detail.

The basic difficulty is that every isolated population is an independent gene pool and evolves independently of what is going on in the main body of the species to which it belongs. For this reason, every peripherally isolated population is potentially an incipient species. Careful analysis of their genetics and the nature of their isolating mechanisms has indeed shown that some of them are on the way to becoming new species, and some of them actually may have already passed this threshold. In areas suitable for such a distribution pattern, particularly in insular regions, every major species is usually surrounded by several populations that have reached the stage of being allospecies, but as far as all of them are concerned, we must make an inference on the basis of all available data and criteria as to how far along they have proceeded on the way to being a separate species. When making this inference, we must be clearly conscious of what we are actually doing. We are studying the available evidence (properties of species populations) to determine whether the species concept (the definition of the concept) is met by the respective populations. The logic of this procedure has been well stated by Simpson [(1961: 69; see also Mayr (1992a: 230)]. This means we do not contend that they are so similar because they belong to the same species, but rather we infer that they belong to the same species because they are so similar. Molecular biology, of course, has given us far more evidence on which to base our conclusions than the purely morphological evidence, which previously was the only evidence available to a taxonomist.

The greatest practical difficulty encountered by the investigator is the occurrence of mosaic evolution. Some populations may acquire repro-ductive isolation but only minimal morphological difference (resulting in sibling species), whereas other populations may acquire conspicuously different morphologies but no isolating mechanisms. Equally, rates of molecular divergence and the acquisition of niche specializations vary independently of the acquisition of reproductive isolation.

Even accepting all the difficulties, it is evident that the endeavor to use all the available evidence to arrive at the correct decision may provide a biologically far more meaningful classification than an arbitrary decision simply based on degree of morphological difference. To be sure, assign-ing populations to biological species on the basis of the set of criteria discussed by Mayr (1969: 181–187) will not eliminate the possibility of an occasional mistake. However, no better method is available to a biologist.

CHRONOSPECIES. Phyletic lineages change over time, some very slowly, others rapidly. In due time the species taxon that represents the phyletic lineage may change sufficiently to be considered a new species taxon, different from its parental species. This is not speciation as properly de-fined but only phyletic evolution, the genetic change within a single lineage; the numbers of species has remained the same. The problem faced by the paleontologist is how to delimit species taxa within a con-tinuous phyletic lineage. This has been attempted by Simpson (1961),

Hennig (1966), and Wiley and Maydem (2000) but quite unsuccessfully. I have presented an analysis of this problem on a previous occasion (Mayr 1988b). Simpson actually had no solution and Hennig's solution was quite arbitrary and unsatisfactory. In the absence of any better criteria the paleontologist is forced to rely on gaps in the fossil record.

ARE THERE VARIANTS OF THE BSC? In the last fifty years a number of species concepts were proposed, claimed to be improvements of the BSC, correcting some of its deficiencies. Frankly, I have not been persuaded that any of these claims is valid.

Simpson's (1961) proposal of an evolutionary species concept accepts the basics of the BSC – that species are isolated reproductive communities. "An evolutionary species is a lineage (an ancestral descendant sequence of populations) evolving separately from others and with its own unitary evolutionary role and tendencies" (1961: 153). However, under this definition, every geographically or chronologically isolated population would qualify as an evolutionary species. Furthermore, it is quite impossible to determine for any population whether it has "its own independent evolutionary role [in the future] and historical tendencies." Neither of these deficiencies is corrected in the recent rewording of Simpson's definition (Wiley and Maydem 2000: 73). Hennig's (1966) species concept is based on the BSC, and he also accepts the specification that a biological species is a reproductive community. However, his definition suffers from the myth that the old species disappears whenever a new species originates. This is indeed true when the new species originates by the splitting of the parental species (*dichopatric speciation*), but it is not true for peripatric speciation in which the parental species may continue more or less unchanged after having given rise (by budding) to a new species (Mayr 2000: 94–95). There is no gap in the parental lineage.

Paterson (1985) proposed a "recognition species concept," which is, however, only a differently worded version of the BSC (Mayr 1996,

2000: 20–22, Raubenheimer and Crowe 1987). It adds nothing to the understanding of the BSC.

The ecological species concept

The so-called ecological species concept (Van Valen 1976), based on the niche occupation of a species, is, for two reasons, not workable. Local populations in almost all the more widespread species differ in their niche occupation. An ecological species definition would require that these populations be called different species, even though, on the basis of all other criteria, it is obvious that they are not. More fatal for the ecological species concept are the trophic species of cichlids (Meyer 1990), which differentiate within a single set of offspring from the same parents. Finally, there are the numerous cases (but none exhaustively analyzed) in which two sympatric species seem to occupy the same niche, in conflict with Gause's rule. All this evidence shows not only how many difficulties an ecological species concept faces but also how unable it is to answer the Darwinian "why?" question for the existence of ecological species.

HOW IMPORTANT IS SPECIES RANK? Willman and Meier (2000: 115–116) think it is all-important. I believe this depends on the situation. In most practical situations, particularly for ecologists and students of behavior working in a local situation, the rank of populations is very important. These workers must know the status of any two populations that either coexist or are in contact with each other. And this is where the BSC permits reaching concrete conclusions more helpfully than any other so-called species concept.

Not many years ago the conservation laws of the United States gave special protection to threatened organisms only when full species were involved. I protested against this interpretation of the law and insisted that specially precious populations should be protected even if they did not have full species status. I applied this argument to the Florida population of the mountain lion (Florida panther) because it was a highly interesting local population, even though not a full species (Mayr and O'Brien 1991). This interpretation was finally accepted by the federal government owing to great pressure by Florida voters. For an ecologist who works on habitat selection of song sparrows in populations in the San Francisco Bay area, it is irrelevant whether the song sparrow of the Aleutian Islands is considered a separate species or not. Too great a stress on species status actually may in certain situations be in conflict with the best interests of conservation.

THE SPECIES TAXON. Species taxon and species concept are often confounded in discussions of the species problem. However, they have strikingly different meanings. The species concept, as explained above, refers to the meaning of species in the workings of nature. The word taxon, on the other hand, refers to a concrete zoological or botanical object, consisting of classifiable populations (or groups of populations) of organisms. The house sparrow (*Passer domesticus*) and the potato (*Solanum tuberosum*) are species taxa. Species taxa are particulars, biopopulations. Being particulars, they can be described and delimited against other species taxa, but they cannot be defined (Ghiselin 1997). In other words a species taxon consists of a group of populations conforming to the definition of a species concept.

Curiously, the word taxon was introduced into systematics only as recently as about 1950. Before then in all situations in which one now uses the word taxon, one had to use words such as category or concept. To speak of a polytypic species category, as I did in 1942, because the word taxon was not yet available, was absurd. The rank of the taxon is given by the category in the Linnaean hierarchy in which it is placed.

A species taxon, consisting of populations, is multidimensional; it consists of allopatric populations. Peripheral populations in space or time may be in an evolutionarily intermediate stage. This is considered an annoying nuisance by the clerical cataloguer but is hailed by the evolutionist as evidence for the action of evolution.

Owing to the incompleteness of the fossil record, only relatively few continuous series of ancestor-descendants are found in which a delimitation of species is impossible. Yet several paleontologists attempted to articulate a species definition that would make a separation of "vertical" species possible. The most frequently mentioned such definition is Simpson's so-called evolutionary species concept, which was discussed above. However, it failed in its basic objective.

THE ONTOLOGICAL STATUS OF THE SPECIES TAXON. There has been a long controversy among philosophers as to the ontological status of the species taxon. Traditionally, and far into the twentieth century, the species was considered by philosophers to be a Platonian class. Naturalists, however, had long appreciated the non-class nature of biological species. To make the invalidity of the class nature of species more visible Ghiselin (1974) and Hull (1976) proposed considering species to be individuals. This called attention to some of the non-class properties of species such as their spatiotemporal localization, their boundedness, their internal cohesion, and their capacity for change (evolution). Even though agreeing that species are not Platonian classes, most biologists and some philosophers were equally unhappy about calling a species an individual when it actually may consist of millions or billions of individual organisms and shows far less cohesion than a single individual.

Therefore, it was proposed by some naturalists that the term population, applied to species for more than one hundred years, be added to the vocabulary of the philosophy of science to designate a phenomenon of nature, biological species, for which neither the term class (set) nor the term individual seems appropriate (Mayr 1988a, Bock 1995). Biological species taxa are biopopulations, not classes. Terminological pluralism is the answer to this diversity.

SUBSPECIES. The acceptance of the BSC produced a tension between the proponents of the Linnaean (typological) species concept and the BSC. When populations were found in a geographically variable biological species that were only minimally ("subspecifically") different from each other, they were ranked as subspecies. From the point of view of information transfer this was a highly useful method. First of all, it prevented the species category from becoming too heterogeneous, containing both highly distinct good biological species and minimally distinct local geographic races. Its second advantage is that it provided at once information on the nearest relative of these geographic variants and their allopatry. When such populations are treated as full species this information is not available. Such information is particularly valuable in larger genera (Mayr and Ashlock 1991: 105). It must be emphasized that the subspecies terminology is a purely taxonomic convenience and has no evolutionary significance. To be sure, some subspecies, particularly geographically isolated ones, may eventually become full species, but most subspecies never achieve that rank.

The BSC, as presented by me in 1942, was based largely on birds (E. Stresemann, B. Rensch, E. Mayr) and insects (K. Jordan, E. Poulton). Speciation in marine echinoids seems to proceed very much as in birds (Mayr 1954), and so it seems in marine bryozoans. There is great need for the study of geographically variable species in the marine organisms.

PSEUDOSPECIES CONCEPTS. In recent years a number of so-called new species concepts were introduced that actually were not new concepts at all but rather new procedures and criteria for the delimitation of species taxa. Their authors ignore the fundamental difference between species concept and species taxon. Bock (1995) has provided a perceptive analysis of the meaning of the terms concept category and taxon.

In the recent monograph on the phylogenetic species concept (Wheeler and Meier 2000), two different phylogenetic "species concepts" are supported. The authors of these two so-called phylogenetic species concepts quite frankly admit that they provide descriptions of particular species taxa. Mishler and Theriot (2000) state "a species is the least inclusive taxon recognized in a formal phylogenetic classification" . . . "Organisms are grouped into species because of evidence of monophyly" . . . "they are the smallest monophyletic groups worthy of formal recognition." In a similar manner, Wheeler and Platnick (2000) say of the phylogenetic species that it is "the smallest aggregation of populations (sexual) or lineages (asexual) diagnosable by a unique combination of character states." Thus morphological difference is their principal species criterion. This is also clear from Platnick's statement "where they exist, well-defined, diagnosable 'subspecies' should simply be called species" (2000: 174). In several other statements they repeat that any population diagnosable by even the slightest clear-cut difference is a species. This is not the definition of a concept and it is not even a satisfactory definition because taxonomists are apt to disagree with each other about what is "diagnosable." It has no relation to the role of species in nature, their "meaning." Hence, it is not a concept.

Wheeler and Platnick admit that acceptance of their species concept would lead to "a drastic increase in the number of species." This, they say, is compatible with a "fundamental goal of all species concepts to discover how many kinds of organisms exist." They do not seem to realize that "different counts of species based on very different definitions of kinds of species" would lead to very different results. Why should one accept that definition of species that leads to the highest totals of numbers of species? The underlying concept of these so-called phylogenetic species concepts is clearly degree of phenotypic difference. It is, for all intents and purposes, a return to the traditional Linnaean species concept.

ASEXUALLY REPRODUCING ORGANISMS (AGAMOSPECIES). The BSC depends on the fact of interbreeding among populations. For this reason the concept is not applicable to organisms that do not have sexual reproduction. Species in asexually (uniparentally) reproducing organisms are rather arbitrarily distinguished on the basis of phenotypic characters. Lateral gene transfer makes the delimitation of many bacterial agamospecies against each other a rather arbitrary matter. These agamospecies have little in common with the traditional species of the eukaryotes. Obviously such *agamospecies* do not answer to the definition of the BSC.

Any endeavor to propose a species definition that is equally applicable to both sexually reproducing and asexual populations misses the basic characteristics of the biological species definition (the protection of harmonious gene pools). Therefore all these attempts have been unsatisfactory. Agamospecies differ from each other by degree of phenotypic difference. They are placed in the Linnaean hierarchy in the category species.

LITERATURE CITED

Bock, W. J. 1986. Species concepts, speciation, and macroevolution. In *Modern Aspects of Species*, K. Iwatsuki, P. H. Raven, and W. J. Bock (eds.). Tokyo: University of Tokyo Press, pp. 31–57.

Bock, W. J. 1995. The species concept versus the species taxon: Their roles in biodiversity analyses and conservation. In *Biodiversity and Evolution*, R. Arai, R. M. Kato, and Y. Doi (eds.). Tokyo: National Science Museum Foundation, pp. 47–72.

Darwin, C. 1887. [Correspondence] (in F. Darwin, *Life and Letters of Charles Darwin, vol. 3*)

Ghiselin, M. T. 1974. A radical solution to the species problem. *Systematic Zoology*, 23:536–544.

Ghiselin, M. T. 1997. *Metaphysics and the Origin of Species*. Albany: State University of New York.

Grant, V. 1994. Evolution of the species concept. *Biologisches Zentralblatt*, 113:401–415.

Hennig, W. 1966. *Phylogenetic Systematics*, translated by D. D. Davis and R. Zangerl. Urbana: University of Illinois Press.

Hull, D. 1976. Are species really individuals? *Systematic Zoology*, 25:174–191.

Kimbel, W. H., and Martin, L. B., eds. 1993. *Species, Species Concepts, and Primate Evolution*. New York: Plenum Press.

Kimbel, W. H., and Rak, Y. 1993, The importance of species taxa in paleoanthropology and an argument for the phylogenetic concept of the species category. In *Species, Species Concepts, and Primate Evolution*, W. H. Kimbel and L. B. Martin (eds.). New York: Plenum Press, pp. 461–484.

Mayr, E. 1942. *Systematics and the Origin of Species*. New York: Columbia University Press.

Mayr, E. 1948. The bearing of the new systematics on genetical problems: The nature of species. In *Advances in Genetics*, Vol. 2. New York: Academic Press, pp. 205–237.

Mayr, E. 1954. Geographical speciation in tropical echinoids. *Evolution*, 8:1–18.

Mayr, E. (ed.) 1957. *The Species Problem*. American Association for the Advancement of Science Publication 50. Washington, DC: American Association for the Advancement of Science.

Mayr, E. 1969. *Principles of Systematic Zoology*. New York: McGraw-Hill.

Mayr, E. 1982. *The Growth of Biological Thought: Diversity, Evolution, and Inheritance*. Cambridge: The Belknap Press of Harvard University Press.

Mayr, E. 1987. The species as category, taxon, and population. In *Histoire du Concept d'Espèce dans les Sciences de la Vie*, J. Roger and J. L. Fischer (eds.). Paris: Fondation Singer-Polignac, pp. 303–320.

Mayr, E. 1988a. Recent historical developments. In *Prospects in Systematics*, D. L. Hawksworth (ed.). The Systematics Association Special Vol. 36. Oxford: Clarendon Press, pp. 31–43.

Mayr, E. 1988b. The why and how of species. *Biology and Philosophy*, 3:431–441.

Mayr, E. 1992a. A local flora and the biological species concept. *American Journal of Botany*, 79:222–238.

Mayr, E. 1992b. Darwin's principle of divergence. *J. Hist. Biol.*, 25:343–359.

Mayr, E. 1996. What is a species and what is not? *Philosophy of Science*, 63(2):262–277.

Mayr, E. 2000. The biological species concept. In *Species Concepts and Phylogenetic Theory*, Q. D. Wheeler and R. Meier (eds.). New York: Columbia University Press, pp. 17–29, 93–100, 161–166.

Mayr, E., and Ashlock, P. D. 1991. *Principles of Systematic Zoology*. New York: McGraw-Hill.

Mayr, E., and Diamond, J. 2001. *The Birds of Northern Melanesia*. New York: Oxford University Press.

Mayr, E., and O'Brien, S. J. 1991. Bureaucratic mischief: recognizing endangered species and subspecies. *Science*, 251:1187–1188.

Mayr, E., and Short, L. L. 1970. *Species Taxa of North American Birds: A Contribution To Comparative Systematics*. Nuttall Ornithological Club Publication 9. Cambridge: Nuttall Ornithological Club.

Meyer, A. 1990. Ecological and evolutionary consequences of the trophic polymorphism in Cichlasoma citrinellum (Pisces: Cichlidae). *Journal of the Linnaean Society*, 39:279–299.

Mill, J. S. 1843. *A System of Logic Ratiocinative and Inductive*.

Mishler, B. D., and Theriot, E. C. 2000. The phylogenetic species concept (*sensu* Mischler and Theriot): Monophyly, apomorphy, and phylogenetic species concepts. In *Species Concepts and Phylogenetic Theory*, Q. D. Wheeler and R. Meier (eds.). New York: Columbia University Press, pp. 44–54.

Paterson, H. E. H. 1985. The recognition concept of species. In *Species and Speciation*, E. S. Verba (ed.). Pretoria, S. Africa: Transvaal Museum, Monograph No. 4, pp. 21–29.

Raubenheimer, D., and Crowe, T. M. 1987. The recognition concept: is it really an alternative? *South African Journal of Science*, 83:530–534.

Roger, J., and Fischer, J. L., eds. 1987. *Histoire du Concept d'Espèce dans les Sciences de la Vie*. Paris: Fondation Singer-Polignac.

Simpson, G. G. 1961. *Principles of Animal Taxonomy*. New York: Columbia University Press.

Sloan, P. 1987. From logical universals to historical individuals: Buffon's idea of biological species. In *Histoire du Concept d'Espèce dans les Sciences de la Vie*, J. Roger and J. L. Fischer (eds.). Paris: Fondation Singer-Polignac, pp. 97–136.

Sonneborn, T. 1975. The *Paramecium aurelia* complex of fourteen sibling species. *Transactions of the American Microscopical Society*, 94:155–178.

Van Valen, L. 1976. Ecological species, multispecies, and oaks. *Taxon*, 25:233–239.

Wheeler, Q. D., and Meier, R., eds. 2000. *Species Concepts and Phylogenetic Theory*. New York: Columbia University Press.

Wheeler, Q. D., and Platnick, N. I., 2000. The phylogenetic species concept (*sensu* Wheeler and Platnick). In *Species Concepts and Phylogenetic Theory*, Q. D. Wheeler and R. Meier (eds.). New York: Columbia University Press, pp. 55–69.

Wiley, E. O., and Maydem, R. L. 2000. The evolutionary species concept. In *Species Concepts and Phylogenetic Theory*, Q. D. Wheeler and R. Meier (eds.). New York: Columbia University Press, pp. 70–89.

Willmann, R., and Meier, R. 2000. A critique from the Hennigian species concept perspective. In *Species Concepts and Phylogenetic Theory*, Q. D. Wheeler and R. Meier (eds.). New York: Columbia University Press, pp. 30, 101–118, 167.

The Origin of Humans

THE STUDY OF THE EVOLUTION OF HUMAN ANCESTORS is at present in considerable turmoil, after a period of some forty or fifty years of relative stability. What is the cause of this current uncertainty? It seems that three different factors are primarily responsible: the recent discovery of five or six new kinds of hominid fossils, a more consistent application of geographic thinking to the ordering of hominid taxa, and the appreciation of the importance of climatic changes for the evolution of hominids. These facts lead to a reevaluation of much of the fossil evidence and to a good deal of healthy and largely unresolved controversy. My aim here is to provide a somewhat speculative report of my own reinterpretation of human prehistory.

An age of typology

Traditionally the study of hominid evolution was fostered by physical anthropologists who had received their training as human anatomists, most often in Germany. Their philosophy was idealistic morphology, the traditional conceptual framework of the anatomists. For them every fossil was a new type and often it was given a new name and, if it seemed to be at all truly distinct, it was even placed in a new genus. Geographic races of *Homo erectus* were described as different genera, *Pithecanthropus* (Java) and *Sinanthropis* (China). One historian in the 1930s listed more than thirty generic and more than one hundred specific names for fossil species of hominids. Swinging Occam's razor quite unmercifully, I cut this down to one or two generic and about five specific names (Mayr 1951). It soon turned out that my lumping had been too drastic, but it was not too far from what the best current hominid classification accepts. However, in recent years there again has been a tendency to return to typology and splitting.

The classical reconstruction

The classical view of mid-twentieth century anthropology of the evolution of the hominids was this: humans originated in Africa and this conclusion is now universally accepted. Indeed, not a single hominid fossil older than 2 million years has been found outside Africa. The early African fossils, somewhat intermediate between chimpanzees and *Homo*, were called Australopithecines, after the first find, the South African *Australopithecus africanus*. Until a few years ago, our concept of the Australopithecines was based exclusively on fossils found, beginning in 1924, in eastern Africa (Ethiopia to South Africa). The ensuing account represents the classical concept of the Australopithecines, as derived from the study of the east-African Australopithecines. As so often in the history

196

of palaeoanthropology, a startling new fossil, *Sahelanthropus*, was recently found in central Africa (Brunet et al. 2002), which requires a rewriting of the story of the Australopithecines. I refrained from doing so, because the next find might require another drastic revision. What I present here is the picture we had before the discovery of *Sahelanthropus*. However, in a short appendix I will describe the diagnostic characteristic features of this oldest hominid fossil.

The Australopithecines, owing to their bipedalism, were considered to be closer to *Homo* than to the chimpanzees, in spite of their small brains. Yet in the total assemblage of their characters they seem to me to be closer to the chimpanzees. For instance, in their habits, in spite of their bipedalism, they were largely arboreal. They had strong sexual dimorphism, the males being at least thirty percent larger than the females. Their brains, about 450 cubic centimeters, were hardly larger than those of chimpanzees and their size hardly increased in the more than 4 million years of their existence. While chimpanzees and gorillas live in the tropical rainforest, the Australopithecines lived in the tree savanna. The time span during which australopithecine fossils are found ranges from about 6 to 2.5 million years. A few late Australopithecines, particularly robust ones, are found as late as 1.9 million years ago.

Two lineages of Australopithecines evolved in eastern Africa, between Ethiopia and South Africa, a gracile one (*afarensis-africanus*) and a robust one (*robustus-boisei*). The two lineages were widely sympatric in South Africa as well as in eastern Africa.

Even though a number of hominid fossils have been found in the time span of 4 to 6 million years ago, when the transition from chimpanzee to *Australopithecus* could be postulated to have occurred, none of them is at the expected half-way stage between the two taxa. There was apparently a good deal of geographic variation at that time and we need a far more thorough analysis of these fossils; we also need more fossils. However, there is no doubt that *Australopithecus* is the intermediate link between chimpanzees and *Homo*. [See below for an evaluation of *Sahelanthropus*.]

This simplified evolutionary history of the hominids had to be revised and expanded for two reasons: an enriched fossil record and a more imaginative interpretation of their environment, particularly climatic changes.

The step from *Australopithecus* to *Homo*

The earliest fossils of *Homo*, *Homo rudolfensis* and *Homo erectus*,[1] are separated from *Australopithecus* by a large, unbridged gap. How can we explain this seeming saltation? Not having any fossils that can serve as missing links, we have to fall back on the time-honored method of historical science, the construction of a historical narrative. We have to make use of every conceivable clue to construct a probable scenario and then test this explanation against all the available evidence. By reconstructing climate and vegetation during the transition period we can actually discover several factors that had been neglected in the past. And we must use Darwin's favorite method: ask questions. Did any climatic change occur at the transition period? What effect would it have on the vegetation? What are the crucial innovations in the anatomy of *Homo*? Why is sexual dimorphism reduced in *Homo*? I will try to answer these questions and a number of additional ones. Readers who are not familiar with the method of historical narratives may say, Why should I believe in any of this, it is nothing but speculation. Yes, you can call it speculation, but this designation ignores that my scenario is based on carefully weighed inferences. And by permitting testing by alternative inferences, it is a most heuristic method. It provides a "most probable" scenario, which suggests new questions one otherwise might not have thought of.

I will not present here a detailed account of these recent developments because I have just published a full account of this history elsewhere

---

[1] I follow those who place *Homo habilis* in the genus *Australopithecus*.

(Mayr 2001). What I present instead is an abridged and somewhat revised treatment.

Changes in climate and vegetation

The decisive motor in human evolution was apparently a series of climatic changes. The Miocene and Pliocene were periods of increasing aridity in Africa. This drought period probably peaked around 2 million years ago. As Africa became more arid, the trees in the tree savanna suffered, more and more of them died, and the tree savanna gradually became a bush savanna. The dying of the trees deprived the australopithecines of their retreat to safety. They were completely defenseless where there were no trees. They were threatened by lions, leopards, hyenas, and wild dogs, all of whom could run faster than they. The australopithecines had no weapons, such as horns or powerful canines, nor the strength to wrestle successfully with any of their potential enemies. Presumably most australopithecines perished in the hundreds of thousands of years of this vegetational turnover. There were two exceptions. Some tree savannas in especially favorable places apparently retained their trees and australopithecines survived here for a while, such as *Australopithecus habilis* and the two robust species (*Paranthropus*). More importantly, some australopithecine populations evolved into *Homo* and became adapted to the bush savanna and its carnivorous inhabitants.

How could australopithecines become adapted to the bush savanna?

For the australopithecines the bush savanna was a rather hostile environment. Lacking the normal defenses (speed, strength, powerful teeth) to cope with the big carnivores, what allowed the australopithecines to live in a treeless environment occupied by lions and hyenas? The only possible answer is ingenuity. The survivors might have thrown rocks, they might have used long poles like some chimpanzees in western Africa, or they might have swung thorn branches and perhaps even used noise-making instruments like drums. Yet, surely fire was their best defense. The discovery of fire was probably the most important step in the evolution of *Homo*. Not being able to sleep in tree nests, they most likely slept at campsites, protected by fires. They were also the first hominids to make flaked stone tools, and it is conceivable that they used sharper flakes to construct lances. The fact is that some australopithecines, now evolving into *Homo*, survived and eventually prospered. The arboreal bipedalism of the australopithecines evolved into the terrestrial bipedalism of *Homo*. The arms shortened and the legs lengthened. But what selection rewarded more than anything else during this shift into this inhospitable new environment, the bush savanna, was ingenuity, brain power. And, indeed, the increase in the size of the brain (from 450 to 700–900 cubic centimeters) was the most conspicuous characteristic of the new genus *Homo*. *Australopithecus* in their physical characteristics (except bipedalism) – small brains, sexual dimorphism, and mode of living – were still chimpanzees. In the long evolution from chimpanzee to *Homo*, the decisive step in hominization was that from *Australopithecus* to *Homo* (see below).

Shifts in the diet

The shift from the rainforest habitat (chimpanzees) to the tree savanna presumably required a considerable change of diet. Trees with soft, tropical fruit were very much rarer in the new habitat and so were plants with lush leaves and soft stems. Obviously the food of the australopithecines in the tree savanna was tougher. Presumably it consisted to a considerable extent of subterranean tubers, but they are tough food. Interestingly, the thickness of the tooth enamel responds apparently rather quickly to selective pressures, and indeed the enamel of the australopithecine teeth (particularly the incisors) is thicker than that of chimpanzees. When the australopithecines became adapted to the bush savanna with its even tougher food, presumably including much tuber food, one would have expected in *Homo* even thicker enamel. But to everyone's astonishment, this is not what one found. *Homo* has thinner enamel than the australopithecines.

How can one explain this seemingly contradictory finding? To what soft food did *Homo* switch? Two (not mutually exclusive) answers to this question have been advanced. According to one, *Homo* adopted carnivory. They were able to take over only partly consumed carcasses of recently killed victims of carnivores, a clearly soft-food item. Fire is the other explanation. It permitted the cooking and roasting of tough plant parts and greatly expanded the amount of available food. A result of his improved diet was a rapid increase in body size. The gracile australopithecines were about four and a half feet tall and weighed about fifty kilograms, while Neanderthals were about five feet five inches tall and weighed about sixty-five kilograms.

Increase in brain size

Brain size was stable in the australopithecines. In more than 2 million years it stayed around 450 cubic centimeters, averaging only very slightly larger than that of chimpanzees. However, the shift to the bush savanna resulted in a near doubling of brain size to 700–900 cubic centimeters in a period of about one-half million years. However, it probably started in an allospecies in central, western, or northern Africa. Eventually, it reached 1,350 cubic centimeters in *Homo sapiens*.

Changes in the newborn

To reach the greatly increased size of the adult brain of *Homo*, the growth of the brain had to be accelerated from the earliest embryonic stages on. But this caused new difficulties during the birth of the infant. Upright stature set a size limit to the mother's birth canal. The head of the newborn could not exceed a certain size and much of the growth of an infant's brain therefore had to be postponed to the postpartum period. In other words, the infant had to be born prematurely. As the growth of the brain was more and more shifted to the postpartum age, the newborn was increasingly backward and helpless. It takes the human newborn about seventeen months to catch up with the agility and independence of the newborn chimpanzee. These "premature" babies require a thick layer of subcutaneous fat, as protection against cooling, and this, simultaneously, makes hair unnecessary or inconvenient. This is why human babies are hairless, compared with the hairy chimpanzee and gorilla infants. In the human infant much of the growth of the brain is postponed until after birth and therefore the size of the brain almost doubles in the first year of life.

Extension of maternal care

As the newborns were more and more premature, selection for increased maternal care became stronger. Fortunately, the mothers now no longer needed their arms for grasping tree branches in an arboreal mode of life (Stanley 1998). As the period of pregnancy lengthened and the mothers also carried their infants until long after birth, greater demands were made on the strength of the females and sexual dimorphism declined. Instead of the males being fifty percent heavier than the females, as in the australopithecines, the difference was reduced in *Homo* to fifteen percent.

The geography of hominid evolution

Classical paleontology and anthropology knew only one dimension, the time dimension. *Australopithecus afarensis* (3.9–2.8 million years ago) from eastern Africa was older than *A. africanus* (2.8–2.3 million years ago) of South Africa. Different species, like *A. afarensis* and *A. africanus*, and *Australopithecus boisei* and *Australopithecus robustus*, were preferably placed in the same phyletic lineage. Where they were located geographically was never emphasized. This ignored the fact that most genera of primates, both in South America and in Africa–Asia, contain superspecies with geographically representative allospecies. In the hominids, *A. afarensis* and *A. africanus* as well as *A. boisei* and *A. robustus* are presumably allospecies. The new Chad species (*Sahelanthropus tchadensis*) is obviously a different allospecies from *A. afarensis* (Brunet et al. 2002). In the few years from 1994 to 2001 no fewer than six new hominid fossils were discovered. Their correct taxonomic assignment will be considerably facilitated if their geographic location is treated as an important taxonomic character.

The incompleteness of fossils

The variety of hominid fossils, particularly the older ones, creates great difficulties for their interpretation. No fewer than four new putative genera of fossil hominids have been described in the last ten years, mostly from single specimens. Will the diagnostic characters of these fossils also be found in future specimens or are they not fully diagnostic?

The difficulties created by the scrappiness of the material are documented by the cranium of *Sahelanthropus tchadensis*. There are no long bones; hence, it is not known whether *S. tchadensis* was bipedal. Therefore this fossil must be compared not only with australopithecines but also with the African apes (chimpanzees and gorillas). On the basis of their genes humans are clearly very closely related to chimpanzees. There-fore, one would therefore expect *S. tchadensis* to be very chimpanzee-like, but they are not. They have many characters one would not have ex-pected in a common ancestor of humans and the chimpanzees. There is the enormous brow ridge, thicker even than in considerably larger-bodied male gorillas. The *foramen magnum*, through which the spinal cord exits, is situated somewhat more forward than in chimpanzees, suggesting some bipedalism. The snout below the nose did not project as much as in a chimpanzee or in *Australopithecus*, thus more resem-bling *Homo*. The incisors are chimpanzee-sized, but the canines are small. The brain case is of the size of a chimpanzee's, but it is lower and narrower. The cheek teeth are larger and thicker-enameled than in chimpanzees, to list the somewhat unexpected mixture of characters of *S. tchadensis*.

S. tchadensis is a beautiful illustration of mosaic evolution. Each feature of the cranium seems to have evolved more or less independently of the others. Much further material is needed before we can understand the early australopithecine evolution. *S. tchadensis* presumably belongs to a different allospecies from the ancestor of *A. afarensis*. The indication of bipedalism and the tougher tooth structure indicate that *S. tchadensis* (6 to 7 million years old) was not an inhabitant of the rainforest but had already acquired some adaptations to the tree savanna.

The next steps

It is rather astonishing how detailed a picture we already have of ho-minid evolution. Molecular biology has irrefutably established *Homo*'s close relationship to the chimpanzees. The ancestral hominids evidently varied geographically but, with no or insufficient material of the differ-ent allospecies being available, it is impossible to infer the connections among the various phyletic lineages. With mosaic evolution apparently rampant, such information might not add much to our understanding of the evolution of the human species.

The gradual evolution from chimpanzee to *Homo* included two major steps. The first one, from the rainforest chimpanzee to the tree savanna semi-chimpanzee *Australopithecus*, may have taken more than one-half million years and may have occurred in several ministeps. It may have taken about 20,000 generations; hence, it proceeded by Darwinian gradualness. The second step from the tree-savanna-inhabiting *Australopithecus* to the bush savanna *Homo* may have been considerably more rapid but it was likewise populational and hence gradual (Wrangham 2001).

The taxonomists had to decide how to classify *Australopithecus*. When discovered in 1924, it was decided, after a long controversy, that *Australopithecus* was closer to *Homo* than to the chimpanzee – in other words, that it was a hominid. This decision was largely based on its bipedal locomotion. One thought the acquisition of upright posture was the most important step in hominid evolution, because it included freeing the hands for tool use. But we have since learned of extensive tool use by chimpanzees and of the complete stasis of brain size in the more than three million years of australopithecine existence. Classical australopithecines had the same brain size as chimpanzees (± 450 cubic centimeters). Indeed, except for the partial bipedalism (but they still lived largely in trees), the australopithecines were chimpanzees. The decisive step to hominid was that from *Australopithecus* to *Homo*.

We do not yet fully understand the nature and variation of the stage in human evolution represented by *Australopithecus* and its relatives. Two further developments are needed. First, we need a very detailed analysis of the "hominid" fossils discovered in the last ten years. So far, most of them have only been given names and a minimal description. And, more importantly, we need more fossils, particularly from parts of Africa beyond eastern and southern Africa. If such fossils are ever discovered, I expect they will require a considerable revision of the classical picture of human evolution.

Appendix

The oldest hominid fossil, *Sahelanthropus*, was discovered in 1997 in the desert region of Chad in central Africa, about 2,500 kilometers distant from the eastern African Rift Valley. It was found to be associated with forty-two taxa of fossil mammals. These fossils, many of them also known from other African localities, permit the dating of the Chad locality at 6–7 million years (upper Miocene). This date is approaching the inferred date when the hominid lineage split from the chimpanzee lineage. One would expect a hominid fossil found at that age level to be intermediate between Australopithecines and chimpanzees. But, to everybody's surprise, this is not what *Sahelanthropus* turned out to be. It is not an Australopithecine with an increase of chimpanzee characters, but a unique mixture of very primitive characters (small brain in a small body) and rather hominid characters (like small canines) and some characters found neither in hominids nor chimpanzees (enormous supraorbital torus). It is an extreme example of mosaic evolution.

How can one explain this combination of characters in a 6- to 7-million-year-old hominid fossil? How does *Sahelanthropus* fit into the hominid phylogeny? The simplest, but by no means necessarily the most likely correct solution would be to consider *Sahelanthropus* one of the australopithecine allospecies. However, it is sufficiently different from the eastern African allospecies of *A. africanus* that it might belong to a different superspecies. With its combination of characters, it is as qualified to be the ancestor of *H. erectus* as is *A. africanus*. I have suggested the possibility on a recent map (Mayr 2001: Fig. 11.3) that *Homo* had descended from a northern or western African (not eastern African) species of australopithecines. However, everything is guesswork until more fossils are found.

LITERATURE CITED

Brunet, M., et al. 2002. A new hominid from the upper Miocene of Chad, Central Africa. *Nature*, 418:145–155.

Mayr, E. 1951. Taxonomic categories in fossil hominids. *Cold Spring Harbor Symposia on Quantitative Biology*, 15:109–118.

Mayr, E. 2001. *What Evolution Is*. New York: Basic Books.

Stanley, S. M. 1998. *Children of the Ice Age: How a Global Catastrophe Allowed Humans to Evolve*. New York: W. H. Freeman.

Wrangham, R. W. 2001. Out of the pan and into the fire: from ape to human. In *Tree of Origin*, F. de Waal (ed.). Cambridge, MA: Harvard University Press, pp. 119–143.

12

Are We Alone in This
Vast Universe?

Humans have asked this question for more than 2,000 years, speculating where some other worlds might be, and the question is still alive. At this moment there are a number of devices in operation listening for signals from extraterrestrials on other planets. This activity is referred to as the search for extraterrestrial intelligence (SETI). To simplify the discussion, I refer to those who believe in the existence of extraterrestrials and who attempt to communicate with them as Setians. Most Setians are physicists or astronomers. The speculations of biologists are more modest. With a few exceptions, they do not ask "are there other human-like creatures on other worlds?" but simply "is there other life somewhere in the universe?". Setians have been running radio telescopes for more than twenty years, not discouraged by the absence of any indications in their recordings that could be interpreted as signals from extraterrestrials. Their opponents believe the evidence opposed to the possibility of success in this endeavor is overwhelming and that it is no longer reasonable to continue the SETI.

What is the reason for the longevity of the argument between the Setians and their opponents?

When reading through the voluminous literature, I was suddenly struck by the realization that two rather different questions were consistently confounded in the controversy:

(1) What is the probability of life elsewhere in the universe?
(2) What is the chance of communicating with extraterrestrials?

What is the probability of life elsewhere in the universe?

The answer to the first question depends on a number of conditions. First of all, we must define what we mean by "life." When laypersons speak of life in the universe, they usually mean human-like extraterrestrials. The late distinguished Harvard astronomer Donald Menzel amused himself with making drawings of the life we might encounter on Mars. All were versions of the human species but some were green, some had some additional extremities, etc. By contrast, when biologists speak of life they think of molecular complexes. This of course involves deciding what life is. I accept a broad definition; life must be able to replicate itself and make use of energy either from the sun or from certain available molecules, like sulfides in the deep sea vents. Such life would consist of bacteria or even simpler molecular aggregates. Biologists who are specialists in this field tend to think that the repeated origin of such life on planets throughout the universe is highly probable. Indeed there are quite a few suggestions in the literature about how a combination of carbon, oxygen, hydrogen, nitrogen, and a few other elements that are widely available in the universe could produce life spontaneously under the proper environmental conditions (temperature, pressure, etc.).

HOW SUITABLE IS THE UNIVERSE FOR LIFE? Setians and their opponents agree that conditions suitable for the origin of life and intelligent life can be found only on planets. Indeed, among the nine solar planets, not only the Earth, but also two other planets (Venus and Mars) have at some stage of their development most likely been suitable for life, presumably a bacteria-like kind of life. If there are billions of planets and a fifth of them (plus or minus) have conditions suitable for life, then surely an availability of planets will not be a problem for the origin of life. And, thus, early Setians took an availability of abundant suitable planets for granted. However, recent studies indicate that the solar planets may be quite exceptional. In all calculations of the probability of life in the universe, it is now necessary to consider the rarity of planets in the universe suitable for life (Burger 2002). Indeed, there are apparently scores of difficult steps between the Big Bang and the origin of a suitable planet.

HOW DIFFICULT WOULD HAVE BEEN THE ORIGIN OF LIFE ON EARTH? Probably not too difficult, considering the abundance of the necessary molecules on the early earth. This conclusion is confirmed by the rapidity with which life appeared on earth after it had become habitable. The proper conditions for life on earth are inferred to have been reached around 3.8 billion years ago. The first fossil bacteria are found in deposits that are 3.5 billion years old. If one postulates that it may have taken about 300 million years for the evolution of modern bacteria from the first origin of life, it would mean that life originated very soon after the earth had become inhabitable.

One might conclude that the origin of life on earth was rather easy because it happened so fast. However, if it were so easy, why did not all sorts of life originate, answering the broad definition of life we have accepted, but only one is found? The genetic code of all organisms now living on earth, down to the simplest bacteria, is, with a few exceptions, identical and this, owing to the arbitrary nature of the code, is convincing evidence that all life now existing on earth had a single origin.

Considering the facility with which life apparently originated on earth, one would postulate that life originated on millions of planets. If so, how does this other life differ from that now found on earth? Did any of it have the potential to develop high intelligence? I am afraid we will never know. And here I touch the fundamental problem of the search for life in the universe. How are we ever to find out whether there is life, in the broadest sense of the word, anywhere else in the universe if such life does not have an electronic civilization enabling it to communicate with us?

Even so, we can now answer our first question. Yes, there is a high probability for the existence of other life, in the broadest sense, some-where else in the universe. Alas, as of this moment, we have no means of finding out whether such life actually exists or existed on a planet beyond the solar system.

What I cannot understand is why the Setians are searching for traces of life with such determination. To find it would be a highly improbable accident. Therefore the search presumably will be unsuccessful. This would prove nothing because life might indeed exist somewhere else but be inaccessible to our search. If life, in the form of some bacteria-like organisms, actually were found unexpectedly, this would tell us very little. Yes, living molecular assemblages might originate occasionally. So what? Is it worth hundreds of million dollars, like the ill-fated recent Mars probe? I doubt it. The money could have been spent far more effectively in researching the rapidly dwindling diversity of the tropical rainforests on earth. But that urgent task is neglected in favor of possibly finding some fossil bacteria on Mars. Should we perhaps organize a search for *terrestrial* intelligence?

What is the chance of communicating with extraterrestrials?

In virtually all the published books and papers on life in the universe the authors begin with a very simple question: Is there life outside of the earth? But then it soon becomes very clear that these Setians could not care less whether some bacteria-like very primitive organisms exist on other planets. What they really want to know is whether there are extraterrestrial organisms with whom we could communicate. But this is of course a very different question from whether life simply exists elsewhere.

The project to get in touch with such organisms, SETI, is primarily supported by physical scientists. Deterministic thinking is quite common in their sciences, in which laws play such an important role. These Setians seem to assume that once life has originated somewhere, in due time it will inexorably evolve into intelligent life. Biologists are not willing to make such a jump. This is why only a few superoptimistic biologists are willing to support the SETI project.

The Setians are up against a formidable problem. How can they determine whether there is life on a far distant planet? They soon realized that, for the time being, there is only one possibility. It is that such life has produced higher organisms that are rather human-like and have developed an electronic civilization. If they have the same urge as we have, to find out whether there is life elsewhere in the universe, they will send out electronic signals to get in touch with us. If we set up large radio telescopes and carefully register all seeming "noises" recorded by this instrumentation, it will by necessity also include whatever signals the extraterrestrials have sent. This search, of course, would discover among the billions of possible forms of life only highly intelligent members of an electronic civilization.

The reasoning of the Setians is based on the assumption that in many places where life originated, it eventually would have led to high intelligence. They assume that natural selection would place such a high premium on intelligence that it would produce it in lots of places in the universe. Carl Sagan said, "Smarter is better." Well, is it really? About 1 billion species of organisms have originated on the earth since the origin of life (Mayr 2001). If Sagan had been right, millions of them should have high intelligence. However, as we know, this faculty emerged on earth only once. Every evolutionist knows how successful natural selection is in producing needed adaptations. Photoreceptor structures (eyes) were acquired independently at least forty times in the animal kingdom. Or to give another example, bioluminescence evidently also contributes much to fitness. As a result, it has independently originated twenty-six times in the living world. We must conclude that if high intelligence had as high a fitness value as eyes or bioluminescence, it would have emerged in numerous lineages of the animal kingdom. Actually it happened only in a single one of the millions of lineages, the hominid line. All other mammals with some amount of intelligence have relatively large brains but not anywhere near the kind of intelligence that would permit such organisms to develop a civilization.

There are numerous ways to demonstrate how utterly improbable the acquisition of high intelligence is. Evolution is branching. Each branch of the evolutionary tree splits into a number of twigs and each of these has the option to produce high intelligence among its offspring. This begins with the hundreds or thousands of species of bacteria, followed by the most primitive early eukaryote organisms that have a nucleus, but most of them are unicellular. From eighty to a hundred phyla of such unicellular eukaryotes (protists) exist, all of them in principle having the option eventually to produce high intelligence. But only a single one actually did. The higher eukaryotes consist of the three kingdoms of plants, fungi, and animals, again all potentially having the choice of producing a lineage with high intelligence, according to Sagan's principle of "smarter is better." But only one of the fifty to eighty phyla of animals produced the vertebrates, hominids, and ultimately *Homo sapiens*. There is not anything deterministic about evolution and the production of high intelligence. Life originated on earth about 3.8 billion years ago. The hominid lineage developed about 300 million years after the origin of life and high intelligence developed less than 300,000 years ago. This shows how infinitesimal the chance was for this ever to happen.

Similar calculations by Diamond (1992) lead to the same finding of an incredibly low probability for the origin of extraterrestrial intelligence.

WOULD THE EXTRATERRESTRIALS BE ABLE TO SEND SIGNALS? Let us assume, for the sake of the argument, that the totally improbable really happened and a large brained human-like organism evolved on some planet. What would be the chance we could communicate with these extraterrestrials? To achieve success, a number of conditions would have to be met. First of all, they would have to have sense organs similar to ours. If their civilization were based on olfactory stimuli or acoustic ones, they would never think of sending electronic messages. This would at once disqualify most of life on earth. Here on earth for several million years we have had troops of highly intelligent hunter-gatherers, but they never would have built radio telescopes because this requires the existence of an electronic civilization. Rudiments of intelligence are found on earth among birds (ravens, parrots) and in a number of orders of mammals (primates, dolphins, elephants, carnivores), but in no case was the intelligence sufficiently highly developed to found civilizations.

Still, we can ask, is every civilization capable of extraterrestrial communication? The answer is clearly no. On earth, since the origin of *Homo sapiens*, we have had already about twenty civilizations, beginning with the Indus and Sumerian civilizations, several others in the near East, the Greek and Roman civilizations, since the fall of Rome the European civilizations, the three American civilizations, and a number of Chinese and Indian civilizations. They came and they went without producing electronic civilizations.

What is particularly characteristic for civilizations is their short life expectancy. Many of them lasted fewer than 1,000 years and none of them has survived several thousand years. If there had been planets with electronic civilizations and these had sent, before 1900, signals to the earth during their short existence, let us say around 1,000, or 1,500, or 1,900 years, no one on earth would have been aware of their signals, because it was before the start of our electronic civilization.

Conclusion

I have here discussed a whole series of factors, each of which makes the possibility of communication with extraterrestrial beings highly improbable. And when one multiplies all these improbabilities with each other, one finds an improbability of astronomical dimensions. The radio telescopes of the Setians reach only a small fraction of the solar galaxy with a limited number of planets. It is for us irrelevant whether there might be life, even intelligent life, somewhere in the infinite universe where it is not accessible to us. And let us always keep in mind that "life in the universe" does not mean humanoids with human intelligence and an electronic civilization, but anything covered by the definition of "life."

LITERATURE CITED

Burger, W. C. 2002. *Perfect Planet, Clever Species*. Amherst, New York: Prometheus Books.

RECENT BOOKS ON EXTRATERRESTRIAL LIFE

Burger, W. C. 2002. *Perfect Planet, Clever Species*. Amherst, NY: Prometheus Books.

Clark, A., and Clark, D. 1999. *Aliens: Can We Make Contact with Extraterrestrial Intelligence?* New York: Fromm International.

Diamond, J. 1992. *The Third Chimpanzee, The Evolution and Future of the Human Animal.* New York: HarperPerennial.

Dick, S. J. 1998. *Life on Other Worlds: The 20th Century Extraterrestrial Life Debate.* Cambridge: Cambridge University Press.

Drake, F., and Sobel, D. 1992. *Is Anyone Out There? The Scientific Search for Extraterrestrial Intelligence.* New York: Delacorte Press.

Koerner, D., and La Vay, S. 2000. *Here Be Dragons. The Scientific Quest for Extraterrestrial Life.* New York: Oxford University Press.

Sullivan, W. 1993. *We Are Not Alone. The Continuing Search For Extraterrestrial Intelligence.* Revised edition. New York: Dutton.

Ward, P. D., and Brownlee, D. 2000. *Rare Earth: Why Complex Life is Uncommon in the Universe.* New York: Copernicus Books.

Glossary

Agamospecies. An asexual species. A species not reproducing sexually. A group of similar individuals reproducing clonally.

Allele. Any of the alternative expressions (states) of a gene.

Allometric growth. Growth pattern in which different parts of the body grow at different rates.

Allopatric. Of populations or species that occupy mutually exclusive geographic ranges.

Allopatric speciation. The evolution of a population into a separate species involving a period of geographic isolation.

Allospecies. A component species of a superspecies. Different allospecies of the same superspecies ordinarily have mutually exclusive geographic ranges.

Anagenesis. Evolutionary change within a single phylogenetic lineage.

Analysis. The dissection of a larger entity or system into its components, which can then be studied more easily. Not to be confused with reduction.

Apomorph. Referring to a newly evolved (=derived) character.

Archaea (Archaebacteria). A group of prokaryotes, inferred to have been important in the origin of eukaryotes.

Asexual reproduction. Any reproduction not involving the fusion of the nuclei of different gametes.

Australopithecine. A bipedal fossil hominid, one of whom gave rise to *Homo*. Intermediate between chimpanzee and *Homo*.

Biological Species Concept (BSC). Defines species as groups of interbreeding natural populations that are reproductively (genetically) isolated from other such groups.

Biota. The flora and fauna of a region.

Bipedal. Walking on two legs.

Budding. The origin of a new side branch of a phyletic lineage by speciation and subsequent entry of this species and its descendents into a new niche or adaptive zone, resulting in a distinct new higher taxon.

Cartesianism. A purely mechanistic philosophy of biology, based on the writings of the French philosopher Descartes.

Chromosome. A deeply staining DNA-containing body in the nucleus of the cell.

Chronospecies. Part of phyletic lineage that differs phenotypically from other sequentially following portions of this lineage.

Cladistics (Cladism). The ordering of species into classes (clades) on the basis of "recency of common descent" – that is, on the basis of the most recent branching point of the inferred phylogeny.

Cladogenesis. The origin and multiplication of species and higher taxa.

Clone. The offspring, derived by asexual reproduction, from a single sexually produced individual.

Deme. A local population of a species; the community of potentially interbreeding individuals of a species at a given locality.

Dendrogram. A diagrammatic drawing in the form of a tree designed to indicate degree of evolutionary relationship.

Determinism. The belief that the endpoint of most processes in inanimate nature is strictly determined by natural laws.

Dichopatric speciation. The origin of a new species by the splitting of a parental species.

Downward classification. Classification from the largest class downward, using the principle of logical division.

Eidos. Plato's term for the unchangeable essence of a natural phenomenon or process.

Electrophoresis. A process of separating different molecules, particularly proteins (polypeptides), according to their differential rates of migration in an electric field.

Emergentism. The view that composite wholes may have properties not evident in their separate components.

Essentialism. A belief that the variation of nature can be reduced to a limited number of basic classes representing constant, sharply delimited types; typological thinking.

Ethology. The science of the comparative study of animal behavior.

Eukaryotes. Organisms with nucleated cells, as, for instance, protists, plants, fungi, and animals; all organisms other than prokaryotes.

Finalism. The concept that every change in the world, particularly the living one, is the result of a cosmic drive that provides it with a purpose; Aristotle's fourth cause.

Gamete. A male or female reproductive cell.

Gause's (exclusion) principle. Two species cannot coexist at the same locality if they have identical ecological requirements.

Geisteswissenschaften. The German word for the humanities.

Gene pool. The total genetic composition of a population.

Genes, regulatory. Genes (DNA) controlling the activity of other genes.

Genetic program. The genotype of every living individual.

Genomics. The comparative study of gene sequences.

Genotype. The genetic constitution of an individual or taxon.

Geographic isolation. The separation of a population (gene pool) by geographic barriers.

Haploid. Having only a single set of chromosomes, as opposed to diploid (two sets of chromosomes), triploid (three sets), etc.

Hierarchy. In classification, the system of ranks that indicates the categorical level of various taxa.

Historical narrative. A proposed explanatory scenario of past events to be tested for its validity.

Holistic. Looking at wholes as more than the sum of their parts with emphasis on properties arising because of organization.

Homologous. A feature in two or more taxa that can be traced back to the same feature in the common ancestor of the taxa.

Isolating mechanisms. Properties that favor breeding with conspecific individuals and inhibit mating with nonconspecific individuals.

Kin selection. Selective advantage due to the altruistic interaction of individuals sharing part of the same genotype, such as siblings, sharing very similar genotypes owing to common descent.

Lebenskraft. An occult force in living organisms responsible for life. Someone believing in such a force is a vitalist. There is no scientific evidence for the existence of such a force.

Levels of selection. When the object of selection belongs simultaneously to two different categories – for example, individual and species. Doubts on the choice of the objects of selection often arose. In the case of species selection, to avoid confusion some authors prefer to use species turnover or species replacement instead of species selection.

Life in the universe. Anything covered by the chosen definition of life, not merely hominids.

Meiosis. Two consecutive special cell divisions in the developing germ cells, characterized by the pairing and segregation of homologous chromosomes; the resulting germ cells have a haploid set of chromosomes.

Mesocosmos. The world from the atoms to the galaxies.

Mimicry. Resemblance in color or structure of members of one species to other species that are distasteful or poisonous.

Mosaic evolution. Different rates of evolutionary change for different structures, organs, and other components of the phenotype in the same group of organisms.

Neo-Darwinism. The original Darwinian paradigm, except for rejecting an inheritance of acquired characters.

Norm of reaction. The range of phenotypes produced by a genotype when interacting with the variable environment.

Ontogeny. The developmental history of an individual organism.

Orthogenesis. The refuted hypothesis that rectilinear trends in evolution are caused by an intrinsic finalistic principle.

Paleontology. The science that deals with the life of past geological periods.

Paradigm. A system of beliefs, values, and symbolic generalizations that, at a given time, dominates a science or branch of science.

Parthenogenesis. The production of offspring from unfertilized eggs.

Peripatric speciation. The origin of a new species by budding from a parental species, which may continue more or less unchanged; achievement of species status by the descendants of a founder population established beyond the periphery of the parental species range.

Phenotype. The complete range of productions of the genotype, including not only structural characteristics, but also physiological and behavioral ones.

Phylogeny, Haeckelian (monophyly). The history of the lines of descent of groups of taxa, based both on the amount of evolutionary change (anagenesis) and on descent from the most recent common ancestor (cladogenesis).

Phylogeny, Hennigian (holophyly). The history of lines of descent of clades, based exclusively on their branching pattern.

Pleiotropy. The capacity of a gene to affect several characters – that is, several aspects of the phenotype.

Polymorphism. The simultaneous occurrence in a population of several discontinuous phenotypes, with the frequency even of the rarest type higher than can be maintained by recurrent mutation.

Polyploidy. The condition in which the number of chromosomes is an integral multiple greater than two of the haploid number.

Polytypic species. A species consisting of several subspecies.

Population thinking. The realization that in biological populations of sexually reproducing organisms every individual is unique.

Program. Coded or prearranged information that guides a process (or behavior) leading it toward the production of a phenotype.

Prokaryotes. Unicellular organisms whose cells have no nucleus.

Punctuated equilibria. Alternation of extremely rapid and normal or slow evolutionary changes in a phyletic lineage.

Ranking. The placement of a taxon in the appropriate level in the taxonomic hierarchy of categories.

Reductionism. The belief that the higher levels of integration of a complex system can be fully explained through a knowledge of the smallest components.

Saltationism. The theory that evolution progresses through "jumps" (discontinuous steps), not gradually.

Scala naturae. The scale of perfection, a belief in upward progression of natural objects, particularly of living ones; a linear progression from the simplest to the most perfect.

Selection, object (unit) of. The entity that is exposed to the process of natural or sexual selection.

SETI. Search for extraterrestrial intelligence.

Setian. One who believes in the existence of intelligent extraterrestrials on other planets and attempts to communicate with them.

Sibling species. Phenotypically very similar or identical populations that are reproductively isolated from each other; a cryptic species.

Speciation. The evolutionary process leading to the multiplication of species.

Speciation, dichopatric. The origin of a new species through the division of a parental species by a geographic, vegetational, or other extrinsic barrier.

Speciation, peripatric. The origin of new species through the modification of peripherally isolated founder populations. (See *Budding.*)

Speciation, sympatric. Speciation without geographic (spatial) isolation; the origin of a new set of isolating mechanisms within a deme.

Species taxon. A population of organisms qualifying for recognition as a species taxon according to a particular species concept.

Stasis. A period in the history of a taxon during which phyletic evolution seems to have been at a standstill.

Subspecies. An aggregate of local populations of a species, inhabiting a geographic subdivision of the range of a species and differing taxonomically from other populations of the species; a subdivision of a polytypic species.

Superspecies. A monophyletic group of closely related and entirely or largely allopatric species that are too distinct to be included in a single species.

Taxon. A monophyletic group of organisms (or of lower taxa) that can be recognized by sharing a definite set of characters.

Teleomatic processes. Processes, the endpoint of which is regulated by natural laws. Aristotle referred to them as caused "by necessity."

Teleonomic processes. Processes that owe their goal-directedness to the influence of an evolved program.

The evolutionary synthesis. The unification of the theories of the population geneticists (anagenesis) with the theories of the naturalists (cladogenesis); the synthesis of the study of genetic change and adaptation with the study of biodiversity and its origins.

Theism. The belief in a personal god.

Typological species concept. The concept that considers a population or group of populations a species when its phenotype is sufficiently different from that of other populations.

Uniformitarianism. The theory of some pre-Darwinian geologists, particularly Charles Lyell, that all changes in the Earth's history are gradual (see *Saltationism*). Being gradual, these changes cannot be considered acts of special creation.

Vitalism. The now thoroughly refuted belief in the existence of an occult invisible force in living organisms responsible for the manifestations of life in any living organism.

Zygote. A fertilized egg; the individual that results from the union of two gametes and their nuclei.

Index